SpringerBriefs in Applied Sciences and Technology

More information about this series at http://www.springer.com/series/8884

Antonio Luque · Alexander Virgil Mellor

Photon Absorption Models in Nanostructured Semiconductor Solar Cells and Devices

Springer

Antonio Luque
Solar Energy Institute
Polytechnic University of Madrid
Madrid
Spain

and

Nanostructured Solar Cells
Ioffe Physical Technical Institute
St Petersburg
Russia

Alexander Virgil Mellor
Solar Energy Institute
Polytechnic University of Madrid
Madrid
Spain

and

Imperial College of Science
and Technology
London
UK

ISSN 2191-530X ISSN 2191-5318 (electronic)
SpringerBriefs in Applied Sciences and Technology
ISBN 978-3-319-14537-2 ISBN 978-3-319-14538-9 (eBook)
DOI 10.1007/978-3-319-14538-9

Library of Congress Control Number: 2014958579

Springer Cham Heidelberg New York Dordrecht London

Printed on acid-free paper

Springer International Publishing AG Switzerland is part of Springer Science+Business Media (www.springer.com)

Preface

This book is oriented to optoelectronic device physicists and engineers who want to enter into the realm of quantum calculations for modeling and development of their devices.

As regards quantum mechanics, device physicists generally know their fundamentals and the extent to which they shape the matters of their study; however, given the complexity of the subject matter, time constraints mean that they only become involved in the simplest quantum calculations.

The common background of solid state device physicists and engineers is related to generation and recombination of electron hole pairs and their transport, but in most cases these mechanisms are used in a phenomenological way and no attempt to model the underlying physical mechanism is undertaken. The same can be said about the energy spectrum of the semiconductors used. Contrastingly, balance analyses are frequent in this background.

Among these phenomena, the energy spectrum and mechanism of absorption are considered as properties of the material and their engineering is not considered. With the advent of nanotechnology, this has ceased to be true; the spectrum and the absorption properties can be engineered. Thus they should no longer be treated as a phenomenological input of the device modeling.

Quantum dots (QDs) and quantum wells (QWs) in semiconductors have a typical size range of 2–20 nm compared to atomic unit cells which are 0.5–0.6 nm. In solid state physics, attempts at applying standard quantum mechanical techniques usually require calculation cells with 10,000 atoms or more, which is about the limit of what is feasible even with high-performance computers.

It is the difference between the two characteristic length scales—the mesoscopic nanostructure and the microscopic crystal unit cell—that makes the problem so large. Luckily, within a good approximation, we can decouple the problem into its mesoscopic and microscopic parts. This reduces the complexity of the problem significantly, making it easier to handle both in terms of computation and analysis. Applying the so-called k·p techniques to nanostructures serves this purpose; this book is devoted to such techniques.

The mathematical tool more commonly applied by device physicists, due to its simplicity and powerful intuitive content, is the effective mass Schrödinger equation associated to a quantum dot or well. The fundamentals and domain of application are explained in the book. Indeed, the book begins by applying this method to a single effective mass equation with a simple square-potential quantum dot with parallelepiped geometry. Despite its simplicity, the example yields a rich set of consequences, which are studied in detail. Furthermore, this simple example will be a crucial component of the more complex situations to be studied later where several mass equations are used.

The studies in this book are centered on zincblende materials. The highest efficiency solar cells available today are made of these materials as well as most LED diodes. These crystallize in the T_d symmetry group, giving a number of general properties to their respective Hamiltonians, which are thoroughly utilized.

In this book we propose and develop a new k·p Hamiltonian, which we name the Empiric k·p Hamiltonian (EKPH). The EKPH uses four bands: the conduction band (*cb*) and three valence sub-bands: the heavy holes (*hh*), light holes (*lh*) and split-off (*so*) sub-bands. Using the EKPH the full energy spectrum introduced by the quantum nanostructure can be obtained in a few seconds using a laptop. The calculation of the photon absorption coefficients, involving over one thousand transitions, may be obtained in 1.5 hours.

In this book the EKPH is primarily applied to quantum dot and quantum well solar cells. The main validation of the EKPH is the reasonable agreement between the results it produces and those measured in real devices. As further validation, reasonable agreement has also been achieved for the GaAs band-to-band photon absorption. These validations are presented in the book. The accuracy of this fast and simple model is sufficient for many applications in device engineering.

As a comparison, the commonly used eight-band Luttinger-Kohn Hamiltonian, modified by Pikus and Bir for the incorporation of strains induced by the nanostructures, is also presented. This so-called LKPB Hamiltonian is applied using a very simple strain field in order to give the constant band offsets used along this book. Using this method, the time taken for calculation of the quantum efficiency of a quantum dot device is about 170 hours and the results are less accurate with respect to the experimental data than those obtained with the EKPH. Of course, the LKPB Hamiltonian is much more accurate than our EKPH as the proper strain field is adopted. Also, times may be shorter if faster calculation techniques are used. Nonetheless, in all we think that our EKPH Hamiltonian serves as a more useful feedback for the development of better optoelectronic devices than the LKPB Hamiltonian.

Acknowledgments

The authors acknowledge the contribution to this work by many of their colleagues and students at the Institute of Solar Energy (IES) of the Polytechnic University of Madrid (UPM) and of the Ioffe Physical-Technical Institute of St. Petersburg. In particular, the contributions of Professor Antonio Martí, of UPM, and that of the Senior Scientist Alexey Vlasov (who authored Sect. 1.2) and the Ph.D. student Aleksandr Panchak (who made renewed calculations of Fig. 2.8) of Ioffe Institute are acknowledged. Also we acknowledge the role of Professor Viacheslav Andreev in providing one of the authors (AL) a great environment to develop part of this book.

They also acknowledge the support of several funding agencies, some of which are as follows:

- To the Spanish Government by the seminal (1997) SUPERCÉLULA 2FD97-0332-C03-01 project and the GENESIS FV grant CSD2006-0004 of the program CONSOLIDER and the NANOGEFES ENE2009-14481-C02-01 project of the Ministry of Science and Innovation, the last two more directly linked to the work presented here.
- To the Regional Government of Madrid by the NUMANCIA grants S-0505/ENE/0310 and S2009/ENE1477, also directly linked to the work presented here.
- To the European Commission (EC) by the important FULLSPECTRUM SES6-CT-2003-502620 integrated project, the IBPOWER 211640 project and the important NGCPV (283798) EC-Japan Cooperative project.
- To the Russian Government by the important Contract no. 14.B25.31.0020 (Ministry of Education and Science, Resolution 220).

Contents

Chapter 1
Introduction

Abstract This chapter starts with a description of quantum wells and quantum dots. Bandgap tuning using quantum wells is described. It is explained how quantum dots can be used to fabricate an intermediate band solar cell: a third generation concept that should allow the current of a solar cell to be increased without reducing its voltage. A description is given of the state of the art of solar cells containing quantum wells and of intermediate band solar cells made with quantum dots. Quantum well solar cells have been able to produce efficiencies similar to their bulk counterparts, but their tunable bandgaps make then attractive. Present intermediate band solar cells only demonstrate a small increase in generated photocurrent and their voltage is reduced, usually leading to cells that present efficiencies that are lower, or only marginally higher, than single gap counterparts. These issues are examined in this chapter and their origin is described. In all cases, the weak light absorption caused by the quantum structures is a main cause of unsatisfactory performance. This is the main topic of this book: the study of the light absorption by the nanostructures. The chapter ends with a description of the whole book.

Keywords Solar cells · Quantum wells · Quantum dots · Zincblende materials

1.1 Concepts in Nanostructured Solar Cells

Together with many others, nanotechnology has invaded the domain of solar cells. Early proposals [1] included the use of quantum wells (QWs) to adjust the bandgap of the solar cells, and the use quantum dots (QDs) as a means of materializing [2] the intermediate band solar cell (IBSC) [3–5]. Additionally, QDs, often deposited from colloidal solutions, are also widely researched for a variety of solar cell concepts which are beyond the scope of this book. Only semiconductor QDs embedded inside other semiconductors, both with zincblende symmetry (such as InAs in GaAs) are the object of this study.

A. Luque and A.V. Mellor, *Photon Absorption Models in Nanostructured Semiconductor Solar Cells and Devices*, SpringerBriefs in Applied Sciences and Technology, DOI 10.1007/978-3-319-14538-9_1

A QW consist of a very thin layer of the QW semiconductor epitaxially grown on the host or barrier semiconductor and covered again by an epitaxial growth of the barrier semiconductor. Usually, the QWs are assembled in arrays. As represented in Fig. 1.1a, the QW solar cell is made by sandwiching a material with QWs between two ordinary semiconductors of n- and p-type respectively. When the QWs are close enough they are sometimes called superlattices.

Other nanostructures are also occasionally used in solar cells with a variety of purposes. We can mention quantum wires formed by long filaments of nano-structure material embedded in the host material, quantum post, or short quantum wires and combinations thereof. These are not further considered in this book.

The adjective quantum refers to that fact that the nanostructure is small enough as to make quantum effects conspicuous. In this respect, the inhomogeneity implied by the QW induces a one-dimensional potential well that leads to eigenfunctions which are the product of three one-dimensional functions each one depending on the variables x, y and z. Taking z as normal to the QW layers, the z-function may be confined within the QW while the x- and y-functions are extended. In contrast, the

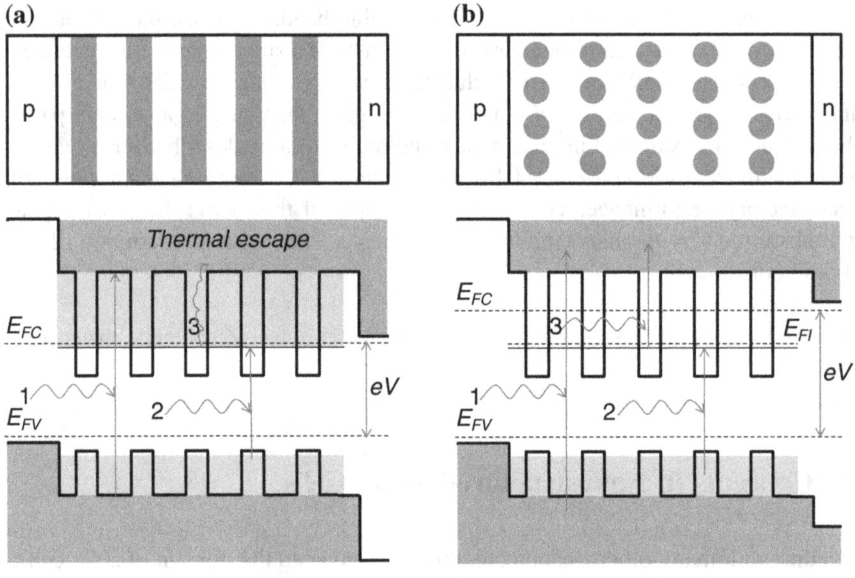

Fig. 1.1 **a** *Above*, the schematic of a QW solar cell and, *below*, its band diagram; besides the pumping of electrons from the VB to the CB by the photons labeled *1*, photons labeled *2* with less energy pump electrons from the VB to the band of QW confined levels. This contributes to the current, since electrons reach the barrier material CB by thermal escape. In this case, the voltage will usually not surpass the bandgap between the VB and the confined levels of the QWs. **b** *Above*, the schematic of a QD solar cell and, *below*, its band diagram. Besides the pumping of electrons from the VB to the CB by absorption of photons labeled *1*, photons labeled *2* with less energy pump electrons from the VB to the IB and photons labeled *3* from the IB to the CB. The voltage can almost reach the CB-VB bandgap. *Grey tones* in the bandgaps represent non-zero density of states: the lighter the lesser

QDs are isolated regions of dot semiconductor inside the barrier semiconductor. These can lead to eigenfunctions that are confined in all dimensions around the QD.

In particular, the potential caused by the QWs is related with the offset of the bands which are well visible in the band diagrams, specifically in that of Fig. 1.1a. As we shall see along the book, the one-dimensional confined wavefunction corresponds to certain discrete energy levels, but because of the extended nature of the other two dimensions, the permitted energies associated to the QW form a continuum that starts in the discrete level and extends upwards as represented in Fig. 1.1a. In the VB, it extends downwards. In the band diagrams in Fig. 1.1 the density of states is (qualitatively) represented by tones of grey, being lighter for lower densities of states. Due to the quantum effects, the energy levels expand inside and outside the QWs.

It is worth noticing that, in the QW, the electron pumped by photon 2 may climb to the CB, where the transport is easy, by absorption of energy from the many phonons and thermal photons available at room temperature. The opposite movement also happens but the thermal equilibrium insures enough density of carriers in the semiconductor CB as to allow for a good transport.

Whereas in the non structured semiconductor the absorption threshold for photons corresponds to the energy of photon 1, in the semiconductor with QWs it corresponds to the energy of photon 2 (in Fig. 1.1a).

The position of the discrete energy level can be regulated by controlling the layer thickness and composition. In this way, these nanostructures are very appropriate to tune the bandgap of the solar cell. This may be useful for multijunction solar cells, where stacks of cells of different bandgaps are used to manufacture very high efficiency solar cells (up to 46 % by autumn 2014).

The IBSC is formed, as represented in Fig. 1.1b, of an intermediate band (IB) material sandwiched between two ordinary semiconductors, one type p and the other type n. In this cell, besides the ordinary pumping with photons 1, electrons may be pumped with photon 2 from the VB to the IB and then with photon 3 from the IB to the CB. In this way two photons of low energy may transport an electron from the VB to the CB, so increasing the cell current. Donors are often added to the QD region to partially fill the IB with electrons and so provide a source for transition 3.

Although the band diagram aspect is almost the same for the QW and the QD solar cell, this is only due to the one dimensional nature of the band diagrams (E-z plots). In reality, the QWs and the QDs have a very different geometrical aspect. The three-dimensional confinement of the QDs leads to isolated energy levels without the continuum of states characteristic of the QWs. In the plot this is represented by the absence of a grey background in the CB potential wells.

Due to the negative effective mass in the VB, the potential pedestals introduced by the nanostructures are equivalent to the wells in the CB. They confine electrons. The reason to draw a light shade for the QDs is that the large effective mass of the heavy holes leads to numerous fully confined states that all together form a quasi-continuum. This means that all the VB states are in strong thermal contact with the extended states of the VB and transitions among them are very frequent causing them to be in thermal equilibrium among themselves.

The voltage of a solar cell is the splitting of the quasi Fermi levels (QFLs) for CB and VB electrons (at their external electrodes, to be precise). In device physics, the electrochemical potentials are called QFLs (E_{FC} and E_{FV} in the drawing). As already said, in QW solar cells there is a relatively large density of states linking the confined-state threshold and the CB. Because of this, all these states are in strong thermal contact with the CB through the interaction with thermal phonons and photons. Therefore, the electrons in these QW states have the same QFL (electrochemical potential) as the CB states. Furthermore, because of the many states, it is very unlikely that this QFL can go much beyond the bottom of the QW threshold of states. Then the voltage is limited to the VB-QW threshold bandgap (see Fig. 1.1a). Contrarily, for the IBSC, there is no continuum of states. Therefore, the IB is (at least ideally) thermally disconnected from the CB. In consequence, the IB has its own QFL (E_{FI}) and nothing prevents the CB and VB QFLs to approach their bands, so permitting a voltage close to the VB-CB bandgap to occur. Because of this the thermodynamic efficiency limit, the IBSC may reach [3] about 63 % versus 41 % for a two-level semiconductor [6], which is what the QW solar cell is.

Unfortunately, this is an ideal situation. Present IBSCs also have some few unwanted states acting as a ladder that thermally connects the IB and the CB, so that the splitting of their QFLs is difficult at room temperature. It is, however, achieved at low temperature and under this condition the cells behave as expected in IBSCs [5].

1.2 Experience with Quantum Well Solar Cells[1]

Research into multi quantum well solar cells (MQW SCs) started in 1990, when Barnham and Duggan suggested the use of QWs for high efficiency solar cells [1]. In general, the main advantage of the MQW SC is that it provides an adjustable bandgap, which is a function of the bandgap widths of the well and barrier materials. The QW area is inserted inside the *i*-region of a *p-i-n* junction, so that the transport through the structure is facilitated due to a built-in electrical field. For solar PV application, the most popular QW system is based on InGaAs wells placed in the GaAs *p-i-n* diode. This structure is capable of increasing the short circuit current of the middle GaAs subcell in triple-junction devices. It enables an increase of the monolithic tandem cell efficiency due a better current matching condition.

First QW SCs were strained InGaAs/GaAs structures grown by a team of Imperial College of London (ICL) [7]. It appeared that the strain, gained in such multiple QW structures, limited the number of QWs, due to misfit dislocations during growth, thus limiting the light absorption. Later, it was shown experimentally by a group from the University of Tokyo (UT) [8] that any residual strain

[1] This section is authored by Alexey Vlasov, of the Ioffe Physical Technical Institute, St Petersburg, Russia.

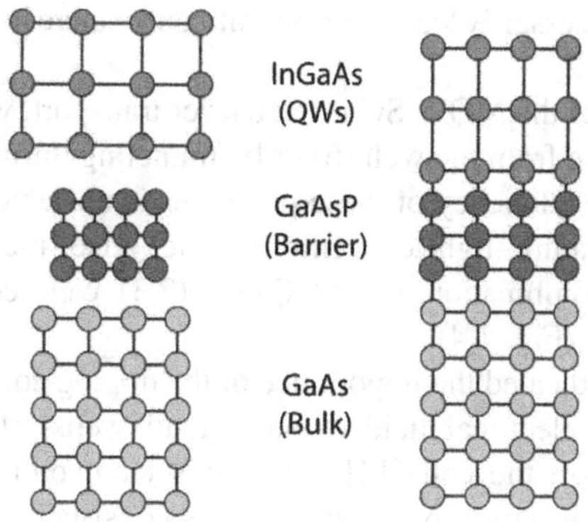

Fig. 1.2 Schematic structure of a strain balanced QW structure. Reprinted with permission. © IOP 2013 [10]

(even a small one) affects the external quantum efficiency (EQE) of the cell. The possible explanation of such behavior is that, in a strained layer, the point defect distribution changes, thus influencing the transport properties of the material.

The approach for the highly efficient solar cells is the use of strain-balanced structures, consisting of InGaAs wells divided by GaAsP barriers (Fig. 1.2). The In and P contents and QW's widths are chosen to minimize the total strain of the system. The first strain-balanced QW cells were also obtained by the ICL team [9, 10]. The cell consisted of 20 MQW's inserted in the GaAs *p-i-n* diode. It was shown that dark I-V curves of the cell were almost unchanged compared to a control GaAs diode, showing an excellent quality of the cell. The EQE spectrum revealed a shoulder in 880–980 nm region with ~20 % efficiency. The I_{SC} value of the QW cell increased a little, although the V_{OC} dropped slightly.

Further attempts were devoted to increasing the QW number for better light absorption. In 2004, Lynch et al. [11] presented a 65 QW SC with an EQE shoulder of 50 % efficiency (which is enough for current matching of the triple GaInP/GaAs-MQW/Ge solar cell) (see Fig. 1.3). At the same time, it was shown that an increase in

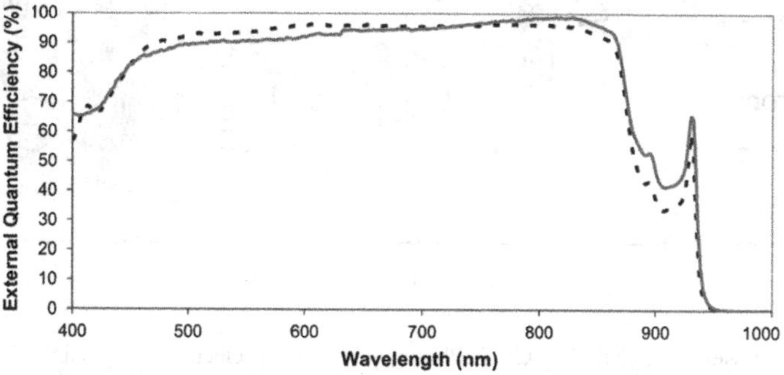

Fig. 1.3 Spectral photoresponse of 50 (*solid line*) and 65 (*dashed line*) MQW solar cells. Reprinted with permission. © Springer 2005 [11]

QW number doesn't necessarily lead to any significant change in V_{OC}, and thus affects mainly the I_{SC} value [12].

The main problem of the MQW SC is the carrier transport, which is realized [13] either by thermal escape from the well (6) or by tunneling through the barrier (5), as shown in Fig. 1.7. The efficiency of these processes is unstable and depends on the electrical field, temperature, light concentration etc. Also, the process of radiative and non-radiative recombination in the QW's (2,3) can reduce the I_{SC} value, affecting the I_{SC} value (Fig. 1.4).

The UT team has indicated the importance of the doping control in the QW SC, a constant value of the electrical field (equal for all wells) should provide better carrier transport through the cell [14]. Also, this team proposed a stepped QW structure for better carrier transport. The QW thus consists of $In_{0.24}Ga_{0.76}As$ wells, $In_{0.13}Ga_{0.87}As$ 1st cladding, GaAs 2nd cladding and GaAsP barriers [15]. The resulting structure revealed better absorption below the GaAs bandgap.

Yet another approach for the strain-balanced QW SC architecture was suggested by the UT team [16]. Normally, the GaAsP barriers contain 10 % phosphorous to keep the barrier bandgap from being too high. In this research the authors introduced superlattices instead of conventional MQW's. In a superlattice, the GaAsP barrier has 43 % phosphorous, which makes it possible to reduce the barrier width down to 3.1 nm, ensuring strain compensation on one hand and good overlapping of wavefunctions on the other. Thus, a band with a lower E_g than that of GaAs is formed. The carrier collection efficiency in such structures did not depend on the electric field. As a result, structures with minimal V_{OC} drop and increased I_{SC} values were presented.

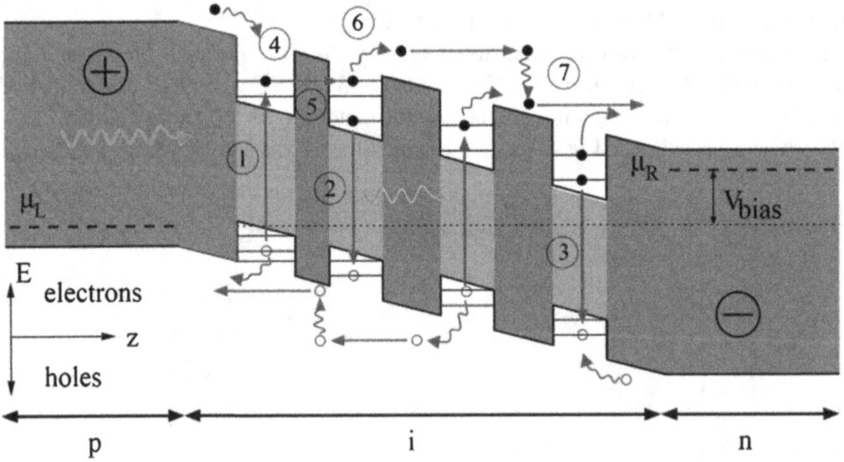

Fig. 1.4 Processes in a MQW SC: *1* Photogeneration of electron-hole pairs. *2* Radiative recombination. *3* Nonradiative recombination (Auger, trap/SRH). *4* Carrier capture. *5* Resonant and non-resonant tunneling. *6* Thermal escape and sweep-out by built in field. *7* Relaxation by inelastic scattering (optical phonons). Reproduced with permission. © SPIE 2010 [13]

Recently the ICL team presented a self-adjusting MQW tandem cell in which both cells (top and bottom) have QW's [17]. In this structure excess carriers in the top InGaAsP/InGaP subcell (e.g. when the blue part of the solar spectrum is enhanced) are captured by the wells and recombine, providing additional radiation for the bottom InGaAs/GaAsP subcell. This process makes the cell less dependent on the solar spectrum, which is very important for terrestrial applications, where the solar spectrum changes hourly and seasonally.

Among other research on MQW SC's, we would like to give attention to the nitrogen based (III-N) system, which has lately become very popular. These materials make use of the potential of the InGaN alloy to overlap the whole solar spectrum (from the 0.65 eV bandgap of InN to the 3.2 eV bandgap of GaN). Also, the reason for so much interest might be the wish to utilize the capability of MOVPE reactors, designed for LED mass-production. The use of QW's in the III-N system seems to be obligatory, due to the strong phase separation of bulk InGaN alloys [18].

1.3 Experience with Intermediate Band Solar Cells

Most of the QD IB solar cell prototypes have been made so far with InAs in a GaAs matrix, grown by molecular beam epitaxy (MBE) in the Stranski-Krastanov mode. The first cell was produced by the Polytechnic University of Madrid in collaboration with the University of Glasgow [19]. However, as is visible in Fig. 1.5, the current increase was negligible and a reduction of voltage was observed, although the contribution of the sub bandgap current is better observed by looking at the quantum efficiency (QE). However this increase is not to be taken straightforwardly as a proof of the IBSC behavior [20].

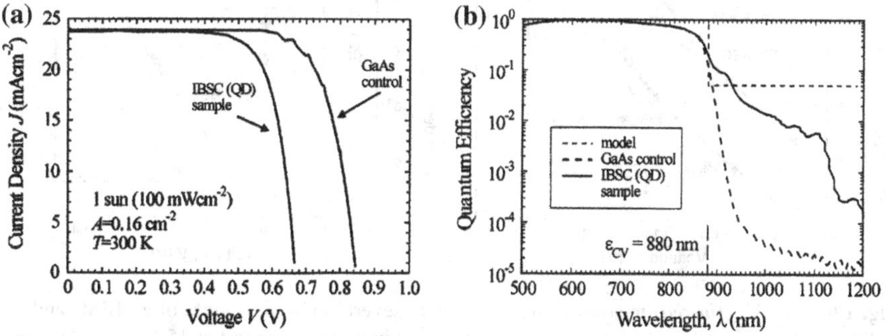

Fig. 1.5 **a** Current-voltage characteristics under one-sun illumination of an IBSC fabricated from InAs QDs and a GaAs control cell. **b** Quantum efficiency of the GaAs control and the QD-IBSC sample. Printed with permission. © 2004, AIP [19]

The deleterious role of the dislocations caused by the growth of the QDs [21] was soon recognized. Actually, the Stranski-Krastanov mode is based on growing a semiconductor of a larger lattice constant (InAs) on one of smaller one (GaAs). The stress is relieved by spontaneous breaking the continuity of the InAs layer, so forming the QDs. However, the structure is still stressed, so resulting in dislocations, which are seriously deleterious for the solar cell operation. Hubbard and coworkers [22] solved this drawback first by insertion thin layers of GaP with a lattice constant still lower than the GaAs, so compensating the effect of the InAs. Actually, they used another growth technique (metal organic chemical vapor deposition-MOCVD) in which the operation with P is easier. The GaP has a larger bandgap too, so that its effect is not visible in the cell operation. Through a very moderate reduction of the voltage and a sensible increase of the current, QD solar cells have been achieved that exceed the efficiency of the test cell of similar characteristics but without QDs [23]. Many other groups have made QD solar cells with slight variations. The best achieved efficiency so far has been 18 %. This was obtained, by a group headed by the Ioffe Institute of St Petersburg [24], through careful cell termination, which is usually neglected by most researchers at this stage of the research.

A strongly simplified IBSC equivalent circuit is given in Fig. 1.6 (inset). It consists of three elementary solar cells linking the VB with the IB, the IB with the CB and a third cell linking the VB and the CB directly. At room temperature, the diode D3 linking the IB and the CB is short circuited by a set of ladder states (to be described later in this book). Consequently, the measured results are characteristic of a cell with the VB-IB bandgap, with less voltage than the reference cell where the bandgap is the VB-CB one. At lower temperatures, the ladder states become less effective and the IBSC voltage converges towards that of the control cell [5, 25, 26]. The voltage of the IBSC cell exceeds that of the reference cell at 20 K (see Fig. 1.6).

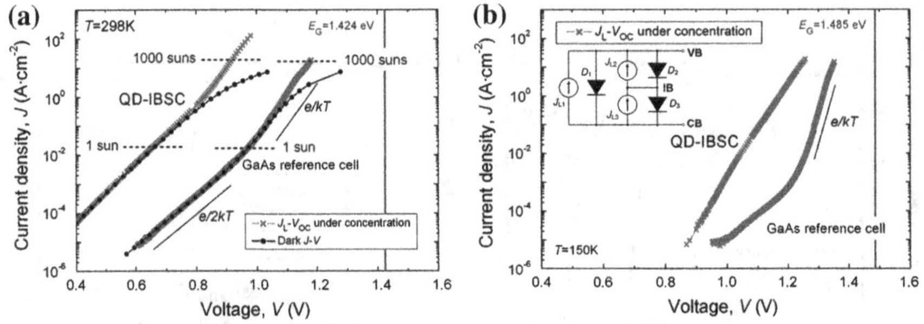

Fig. 1.6 **a** J_{sc}-V_{oc} (*in red*) measured characteristic at several light irradiances of an IBSC and an ordinary reference cell made for comparison at room temperature (**a**) and at 150 K (**b**). At room temperature, the dark current is also drawn (*in black*). Under the superposition principle both *curves* should be the same, but series resistance and other effects make them different. A simplified equivalent circuit is also drawn at the *inset*. Printed with permission. © 2012, Elsevier [25]

Sub bandgap current has been observed although it is too weak. The two photon current associated to the mechanism 2 and 3 of Fig. 1.1b have also been unequivocally observed [27]. The splitting of the Fermi level into three QFLs has also been observed [28]. However, practical IBSCs are not yet available. The optical properties of this structure are a key in understanding the IBSC. This book is devoted to this issue.

1.4 The Zincblende Materials

Most of the III-V semiconductors crystallize in the zincblende structure within the cubic system. This is at least the case for AlP, AlAs, AlSb (with some doubts), GaP, GaAs, GaSb, InP, InAs and InSb, This structure consists of two face-centered cubic lattices, one formed by atoms of the Column IIIA and the other by atoms of the Column VA of the periodic table. Several binary compounds may intermix and form solid solutions that often conserve the same basic structure. We present in Fig. 1.7 the position of the atoms in the zinblende conventional unit cell.

The V type atoms are located in the center of a tetrahedron in whose vertices III atoms are located. This is visible in the figure. The same can be said of the III type atoms, which are also located in the center of a tetrahedron with V atoms at the vertices. Therefore, they have a tetrahedral coordination with four atoms of different nature surrounding each atom. This gives a characteristic band structure with one conduction band and three valence bands, the latter being called the heavy holes, light holes and split off bands. The split off band is at deeper energies in the valence band. This will be seen along this book in closer detail.

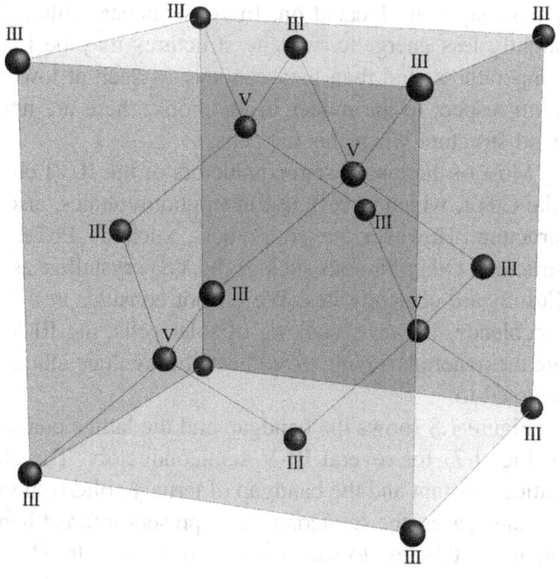

Fig. 1.7 Zincblende lattice. It is formed by two face centered lattices each one displaced by a vector (1/4,1/4,1/4), where the side of the containing cube is taken to be unity. *III* and *V* represent the sites of atoms of the corresponding column

Assuming the length of the containing cube side in Fig. 1.7 to be unity, the III atoms are at the positions (0,0,0), (1,0,0), (0,1,0), (0,0,1), (1,1,0), (1,0,1), (0,1,1), (1,1,1), (0,1/2,1/2), (1/2,0,1/2), (1/2,1/2,0), (1,1/2,1/2), (1/2,1,1/2), (1/2,1/2,1) forming a full face-centered cube unit cell. The V atoms are in (1/4,1/4,1/4), (3/4,3,4,1/4), (3/4,1/4,3/4), (3/4,3/4,1/4), although outside this unit cell they form a full face centered cube lattice. The basis vectors are (0,1/2,1,2), (1/2,0,1/2), (1/2, 1/2,0), which, along with (0,0,0), form the primitive cell. This cell contains two atoms, one III atom formed by 1/8 of the eight atoms in the corners, and one inner V atom. The compact planes are perpendicular to the principal diagonal. Such planes are formed of III atoms situated in a hexagonal array alternated with similar planes formed of V atoms.

This structure is identical to the diamond structure shown also by silicon and germanium when the two different atoms of the binary semiconductors are the same.

The zincblende semiconductors belong to the T_d symmetry group. The structure is invariant under ternary rotations along the 4 principal diagonals and three binary rotations around axes passing by the center of the faces. An improper rotation is a rotation followed of a reflection by a plane perpendicular to the rotation axis. The T_d is also invariant under six improper quaternary rotations around face-diagonal axes and under six reflections around face-diagonal planes. As we shall see this symmetry structure has an indirect influence on the results to be presented in the following chapters, whereas the nanostructure symmetry has a direct influence.

This zincblende structure is close to the wurtzite structure in which the III and V close hexagonal packing structures of the atoms are interpenetrated in such a way that the all atoms are again tetrahedrally coordinated with atoms of different nature. The difference between the two structures is similar to that between the face centered cube and the close hexagonal packing: both structures are the closest way of packing equal spheres and only differ in which underlying interstices the spheres of a third layer are located on. In consequence, although the zincblende structure is slightly less energetic, wurzite structures may be found in these binaries at high temperatures, and then they can stay as such at low temperature if frozen rapidly. With respect to the matter in this book, there are not drastic modifications in the band structure when this happens.

Most of the binary semiconductors of the II-VI columns of the periodic system, like CdTe, which is very useful in photovoltaics, also crystallize in the zincblende structure. However, several others, such as PbTe, crystallizes in the rock-salt structure. Other binaries such as the VS_2 crystallize as spinels. Ternaries such as the $CuInS_2$ are chalcopyrites. We do not consider in this paper structures different to zincblende. However, talking of solar cells, the III-V binaries or alloyed ternaries are the materials having given best photovoltaic efficiency and are those this book is devoted to.

Figure 1.8 shows the bandgap and the lattice constant (the size of the cube sides in Fig. 1.7) for several III-V semiconductors. The chart also shows lines for the lattice constant and the bandgap of ternary solid solutions of the binary compounds. In some cases, the semiconductors present indirect bandgaps; the application of the theory of this text to such cases may be not straightforward.

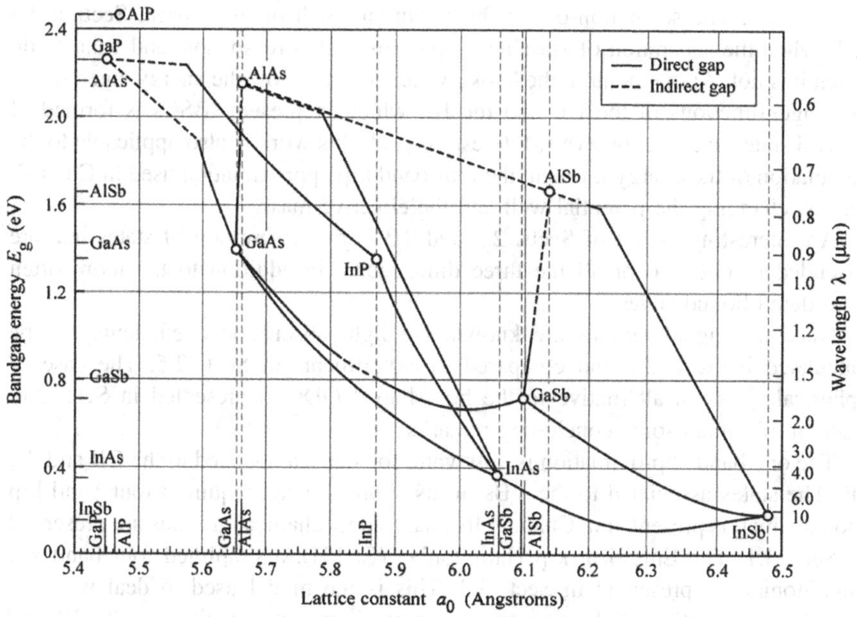

Fig. 1.8 Chart presenting the lattice constant (the size of the cube side of Fig. 1.7) and the bandgap of several the III-V semiconductors. These semiconductors may form alloys represented along the *lines*. In some cases, the bandgap is indirect. This case is not considered in this text

Not all the ternary compounds are possible, although the miscibility is high. Some ternaries may segregate into binary compounds saturated with the foreign element.

This chart is valid for bulk homogeneous materials, nanostructured materials may be strained by the presence of a neighboring different material, and both the lattice constant and the bandgap change value consequently. A detailed study of many III-V semiconductors, most of them crystallizing with the zincblende lattice, can be found in Ref. [29].

1.5 Book Structure

This book is organized as follows: after this introduction, we introduce in Chap. 2 the single band effective-mass equation. This will be introductory to the four band effective-mass equations to be introduced later in Chap. 3. The quantum mechanical grounds of the single band effective-mass equation are described in Sect. 2.1. The square-potential box-shaped model of the QDs and its solution by separation of variables is then solved in Sect. 2.2. However, the separation-of-variables solution is not exact. An exact solution, within the approximations implied by the one-band approximation and the square-potential box-shaped model of the QDs, is presented in Sect. 2.3. The separation-of-variables solutions will often be used. Section 2.3 tells when the separation of variables approximation is reasonable and what to do when it is not. At that point in the book, we are able to know the energy eigenvalues and eigenfunctions of the CB and the IB, which in present IBSCs is formed of several states and energy levels. Interestingly, all this work is also applicable to the calculation of the energy levels in the four band k·p approximation used in Chap. 3, by just changing the potential well and the effective mass.

An interesting aspect of Sects. 2.2 and 2.3 is the description of states that are extended in one two or all the three dimensions, in addition to the more often considered bound states.

Once the eigenfunctions are known, the light absorption coefficients can be calculated in Sect. 2.4 and compared to experiments in Sect. 2.5. The case of spherical QDs, an alternative to the box-shaped QDs, is presented in Sect. 2.6. Section 2.7 offers some concluding remarks.

The one band approximation is only valid for states associated to the CB and the IB. The states associated to the VBs, in its simplest way, require a four band k·p model which is presented in Chap. 3. Its quantum mechanical grounds are presented in Sect. 3.1. The Empirical k·p Hamiltonian (EKPH), a simplified four band k·p Hamiltonian, is presented in Sect. 3.2. This is the model used to deal with the complete description of the nanostructured semiconductor, including CB, IB and VB states. It includes the complete energy eigenvalues spectrum and the corresponding eigenfunctions for the box shaped Hamiltonian as well as the absorption coefficients for interband transitions (transitions between VB states and CB states) and makes extensive use of the theory developed in Chap. 2. Based on the model, the quantum efficiency is compared with actual measured values in an exemplary quantum dot intermediate band solar cell.

An introduction to the group theory applied to the nanostructured semiconductors in order to reduce the calculation burden is presented in Sect. 3.3. Interband transitions (transitions between VB and IB/CB states) between extended and bound wavefunctions are presented in Sect. 3.4. For the intraband transitions, this had already been calculated in Sect. 2.2.4.

Concluding remarks are given in Sect. 3.5. In summary, the absorption coefficients between the levels inside the VB and the several levels forming the IB, and the levels inside the VB and the levels inside the CB are calculated in Chap. 3.

The purpose of Chap. 4 is to write the balance equations of the transitions involving photon absorptions and emissions between these levels or group of levels. Only transitions between bound sates are considered in this Chapter because, as proven in Chap. 3, these are the most important.

An important aspect taken into account in Chap. 4 is that the source of photons is not only the sun but also the thermal photons emitted or absorbed by the surroundings. This is found to be a source of transitions that in certain cases may be dominant.

Chapter 4 starts with a brief introductory Sect. 4.1, that introduces the characteristics and notation on the exemplary cell used to develop the theory. Then Sect. 4.2 develops the detailed balance model for the regime of weak photon absorption. This is the regime in which the present QD IBSCs operate. In Sect. 4.3, all the absorption coefficients relevant for the detailed balance study are calculated, following Chap. 3, and are presented for an exemplary QD IBSC. Section 4.4 calculates the quantum efficiency for different room temperatures permitting a deep insight of the experimental operation of the exemplary cell. For instance, the observed elimination of the sub-bandgap quantum efficiency by decreasing the temperature is reproduced and the Arrhenius law of this elimination is calculated for the first absorption peak in very good agreement with the observed value. Section 4. 5 calculates the IV curve in absence of non-radiative recombination. This is indeed an optimistic assumption but represents the limit of performance of the solar cell and introduces good insight on physical aspects of the cell operation. However, that main reason why the exemplary prototype and most of the cells manufactured so far present poor efficiency increase with respect to ordinary cells, if any, is the low number of QDs introduced. This leads to a weak additional photon absorption and jeopardizes any attempt of increasing substantially the efficiency. Section 4.6 modifies the detailed balance model to include a high absorption regime. This high absorption is to be achieved by increasing the number of QDs and by light confinement. Both aspects have different roles in the solar cell behaviour; increasing the number of QDs improves the photo-generated current but also the radiative recombination, whereas light confinement improves the photo-generated current without increasing radiative recombination. In this chapter, the conditions to render practical an InAs/GaAs IBSC are presented. This theory can be applied to other materials that may be more adequate for manufacturing IBSCs.

Concluding remarks are given in Sect. 4.7.

Chapter 5 contains the application of the EKPH to the case of quantum well solar cells and compares the calculated and measured quantum efficiencies of an experimental quantum well solar cell prototype with reasonable agreement. Chapter 6 applies the EKPH to the case of a bulk semiconductor, in this case GaAs. The absorption coefficient calculated is in good agreement with the classical measurements of this parameter. This confirms the adequacy of the EKPH for the study of absorption coefficients and energy spectra in solar cells and other optoelectronic devices. Sections of concluding remarks are offered in both chapters.

In Chapter 7, the Luttinger-Kohn (LK) Hamiltonian is presented with the Pikus-Bir (PB) modification for strained semiconductors. These are the ones commonly used by theoreticians in solid state physics to undertake the analysis that in this book is made with the EKPH. The purpose of this chapter is to compare the LKPB Hamiltonian with the EKPH. This requires a deeper insight of the spin concept. Section 7.1 describes the spinors and Sect. 7.2 describes the spin-orbit coupling in zincblende materials. Section 7.3 describes the LK Hamiltonian and how to calculate it. In Sect. 7.4, a fitting of the parameters used in the LK Hamiltonian to obtain the experimental effective masses of the InAs is presented.

The introduction of the QDs produces strained lattices. A very simple strain model able to produce squared band offsets similar to those studied in the EKPH is introduced in Sect. 7.5 and the fitting to recover the experimental effective masses and bandgaps are also described. Section 7.6 develops the LKPB Hamiltonian, which is an eight band k·p Hamiltonian, and calculates eigenvalues and absorption coefficients for our exemplary QD-IBSC. The conceptual similarities are very strong but the time necessary for these calculations is much bigger. The comparison with the measured quantum efficiency is reasonable but not better than with the EKPH. This is to be attributed to the approximations we have made to compare the LKPB Hamiltonian with the EKPH rather than to the LKPB Hamiltonian itself. Section 7.7 develops an approximation of the LKPB Hamiltonian that renders it a four band Hamiltonian, like the EKPH. In this case the similarities are very strong and the time needed is also similar, but there are important differences and, in all, the EKPH seems closer to the experimental results.

Conclusions are presented in Sect. 7.8.

Summarizing, the book offers methods for calculating the energy spectrum, the absorption coefficients and the detailed balance of the traffic of electrons when photons are absorbed and emitted in optoelectronic devices and in particular in solar cells.

A new Hamiltonian, the EKPH, which has been developed by the authors, is presented in this book and forms fundamental part of it. It is very simplified and tries to remain close to the device developer and to his/her expected level of theoretical knowledge, so that the developer may use it to feedback the development tasks. It is expected that the book provides what is necessary to this goal.

References

1. Barnham KWJ, Duggan G (1990) A new approach to high-efficiency multi-band-gap solar-cells. J Appl Phys 67(7):3490–3493
2. Martí A, Cuadra L, Luque A (2000) Quantum dot intermediate band solar cell. In: Paper presented at the Proceedings of 28th IEEE photovoltaics specialists conference, New York
3. Luque A, Martí A (1997) Increasing the efficiency of ideal solar cells by photon induced transitions at intermediate levels. Phys Rev Lett 78(26):5014–5017
4. Luque A, Martí A (2001) A metallic intermediate band high efficiency solar cell. Prog Photovoltaics: Res Appl 9(2):73–86
5. Luque A, Marti A, Stanley C (2012) Understanding intermediate-band solar cells. Nat Photonics 6(3):146–152. doi:10.1038/nphoton.2012.1
6. Shockley W, Queisser HJ (1961) Detailed balance limit of efficiency of p-n junction solar cells. J Appl Phys 32(3):510–519
7. Barnham K, Ballard I, Barnes J, Connolly J, Griffin P, Kluftinger B, Nelson J, Tsui E, Zachariou A (1997) Quantum well solar cells. Appl Surf Sci 113:722–733. doi:10.1016/S0169-4332(96)00876-8
8. Sodabanlu H, Ma SJ, Watanabe K, Sugiyama M, Nakano Y (2013) Impact of strain accumulation on InGaAs/GaAsP multiple-quantum-well solar cells: direct correlation between in situ strain measurement and cell performances. Jpn J Appl Phys 51(10). doi:10.1143/Jjap.51.10nd16 (Artn 10nd16)
9. Barnham KWJ, Ballard I, Connolly JG, Ekins-Daukes N, Kluftinger BG, Nelson J, Rohr C, Mazzer M (2000) Recent results on quantum well solar cells. J Mater Sci-Mater El 11(7):531–536. doi:10.1023/A:1026587616640

10. Ekins-Daukes NJ, Barnham KWJ, Connolly JP, Roberts JS, Clark JC, Hill G, Mazzer M (1999) Strain-balanced GaAsP/InGaAs quantum well solar cells. Appl Phys Lett 75(26):4195–4197
11. Lynch MC, Ballard IM, Bushnell DB, Connolly JP, Johnson DC, Tibbits TND, Barnham KWJ, Ekins-Daukes NJ, Roberts JS, Hill G, Airey R, Mazzer M (2005) Spectral response and I-V characteristics of large well number multi quantum well solar cells. J Mater Sci 40 (6):1445–1449. doi:10.1007/s10853-005-0580-4
12. Bushnell DB, Tibbits TND, Barnham KWJ, Connolly JP, Mazzer M, Ekins-Daukes NJ, Roberts JS, Hill G, Airey R (2005) Effect of well number on the performance of quantum-well solar cells. J Appl Phys 97(12). doi:10.1063/1.1946908 (Artn 124908)
13. Aeberhard U (2010) Microscopic theory and numerical simulation of quantum well solar cells. Phys Simul Optoelectron Devices Xviii 7597. doi:10.1117/12.845478 (759702)
14. Fujii H, Wang YP, Watanabe K, Sugiyama M, Nakano Y (2013) Compensation doping in InGaAs/GaAsP multiple quantum well solar cells for efficient carrier transport and improved cell performance. J Appl Phys 114(10). doi:10.1063/1.4820396 (Unsp 103101)
15. Wen Y, Wang YP, Nakano Y (2012) Suppressed indium diffusion and enhanced absorption in InGaAs/GaAsP stepped quantum well solar cell. Appl Phys Lett 100(5). doi:10.1063/1.3681785 (Artn 053902)
16. Sugiyama M, Wang YP, Fujii H, Sodabanlu H, Watanabe K, Nakano Y (2013) A quantum-well superlattice solar cell for enhanced current output and minimized drop in open-circuit voltage under sunlight concentration. J Phys D Appl Phys 46(2). doi:10.1088/0022-3727/46/2/024001 (Artn 024001)
17. Ekins-Daukes NJ, Lee KH, Hirst L, Chan A, Fuhrer M, Adams J, Browne B, Barnham KWJ, Stavrinou P, Connolly J, Roberts JS, Stevens B, Airey R, Kennedy K (2013) Controlling radiative loss in quantum well solar cells. J Phys D Appl Phys 46(26). doi:10.1088/0022-3727/46/26/264007 (Artn 264007)
18. Jani O, Ferguson I, Honsberg C, Kurtz S (2007) Design and characterization of GaN/InGaN solar cells. Appl Phys Lett 91(13). doi:10.1063/1.2793180 (Artn 132117)
19. Luque A, Martí A, Stanley C, López N, Cuadra L, Zhou D, Mc-Kee A (2004) General equivalent circuit for intermediate band devices: potentials, currents and electroluminescence. J Appl Phys 96(1):903–909
20. Marti A, Antolin E, Linares PG, Luque A (2012) Understanding experimental characterization of intermediate band solar cells. J Mater Chem 22(43):22832–22839. doi:10.1039/c2jm33757f
21. Marti A, Lopez N, Antolin E, Canovas E, Luque A, Stanley CR, Farmer CD, Diaz P (2007) Emitter degradation in quantum dot intermediate band solar cells. Appl Phys Lett 90:233510–233513
22. Hubbard SM, Cress CD, Bailey CG, Raffaelle RP, Bailey SG, Wilt DM (2008) Effect of strain compensation on quantum dot enhanced GaAs solar cells. Appl Phys Lett 92(12):123512 (Artn 123512)
23. Bailey CG, Forbes DV, Polly SJ, Bittner ZS, Dai Y, Mackos C, Raffaelle RP, Hubbard SM (2012) Open-circuit voltage improvement of InAs/GaAs quantum-dot solar cells using reduced InAs coverage. IEEE J Photovoltaics. doi:10.1109/JPHOTOV.2012.2189047
24. Blokhin SA, Sakharov AV, Nadtochy AM, Pauysov AS, Maximov MV, Ledentsov NN, Kovsh AR, Mikhrin SS, Lantratov VM, Mintairov SA, Kaluzhniy NA, Shvarts MZ (2009) AlGaAs/GaAs photovoltaic cells with an array of InGaAs QDs. Semiconductors 43(4):514–518
25. Linares PG, Marti A, Antolin E, Farmer CD, Ramiro I, Stanley CR, Luque A (2012) Voltage recovery in intermediate band solar cells. Sol Energy Mater Sol Cells 98:240–244. doi:10.1016/j.solmat.2011.11.015
26. Luque A, Linares PG, Antolín E, Ramiro I, Farmer CD, Hernández E, Tobías I, Stanley CR, Martí A (2012) Understanding the operation of quantum dot intermediate band solar cells. J Appl Phys 111:044502. doi:10.1063/1.3684968

27. Marti A, Antolin E, Stanley CR, Farmer CD, Lopez N, Diaz P, Canovas E, Linares PG, Luque A (2006) Production of photocurrent due to intermediate-to-conduction-band transitions: a demonstration of a key operating principle of the intermediate-band solar cell. Phys Rev Lett 97(24):247701–247704
28. Luque A, Marti A, Lopez N, Antolin E, Canovas E, Stanley C, Farmer C, Caballero LJ, Cuadra L, Balenzategui JL (2005) Experimental analysis of the quasi-Fermi level split in quantum dot intermediate-band solar cells. Appl Phys Lett 87(8):083503–083505
29. Vurgaftman I, Meyer JR, Ram-Mohan LR (2001) Band parameters for III-V compound semiconductors and their alloys. J Appl Phys 89(11):5815–5875

Chapter 2
Single Band Effective Mass Equation and Envolvent Functions

Abstract The single-band effective mass Schrödinger equation to calculate the envelope functions is described and its grounds are shown. These envelope functions are used to multiply periodic part of the Bloch functions to obtain approximate eigenfunctions of the Hamiltonian of a nanostructured semiconductor. The Bloch functions, which are the product of a periodic function and a plane wave, constitute the exact solution of a homogeneous semiconductor; they are taken as a basis to represent the nanostructured Hamiltonian. The conditions that make possible the use of this single band effective mass Schrödinger equation are explained. The method is applied to the calculation of the energy spectrum of quantum dots for wavefunctions belonging to the conduction band. A box-shaped model of the quantum dots is adopted for this task. The results show the existence of energy levels detached from this band as well as eigenfunctions bound totally or partially around the quantum dot. The absorption coefficients of photons in the nanostructured semiconductor, which is our ultimate goal, are calculated. The case of spherical quantum dots is also considered.

Keywords Solar cells · Quantum calculations · Quantum dots · Energy spectrum · Absorption coefficients

The easiest way to approach to the behavior of electrons in semiconductors is the use of effective mass equations. This is presented in this chapter for the specific cases of nanostructured semiconductors. The approximate treatment presented here applies to the case that the nanostructure has a mesoscopic size, much larger than the size of the semiconductor microscopic structure, that is, where the individual atoms are located.

We shall see that the effective mass equation is only of application when we are dealing with electrons in a single semiconductor band, namely the conduction band. Therefore, the bipolar behavior, crucial for the understanding of most semiconductor devices, in which both electrons and holes enter into the game—or in other words, when electrons in the CB and in the VB are to be considered—cannot be explained with the resources of the simple effective mass treatment studied in this chapter. They will be properly studied in the following chapter. Nevertheless, the

A. Luque and A.V. Mellor, *Photon Absorption Models in Nanostructured Semiconductor Solar Cells and Devices*, SpringerBriefs in Applied Sciences and Technology, DOI 10.1007/978-3-319-14538-9_2

detailed study of the simple effective mass equation, besides its intrinsic interest, is a needed step towards the study in the following chapter.

Furthermore, in nanostructured materials, another crucial concept arises which is that of envelope functions. This concept is studied in this chapter for the first time in this book, although it will also be the subject of other chapters for more complicated cases.

2.1 Quantum Mechanical Grounds

We present in this section the quantum mechanical grounds that lead to the use of the effective mass equation and also the envelope functions. This section is mainly inspired in Datta [1, Chap. 6], whose full reading is recommended.

As we have indicated, the nanostructures studied here are of mesoscopic size, in that their dimensions are much larger than the interatomic spacing of their constituent semiconductors. Because of this, we present first the integral factorization rule that will be used in approximate calculations along this book.

2.1.1 The Integral Factorization Rule

Let us consider first that $f(\boldsymbol{r})$ is a slowly varying function (it is assumed to vary in the mesoscopic range) and $g(\boldsymbol{r})$ is periodic with the short period of the crystalline unit cell $\Omega_{cell,0}$ (the crystal is a repetition of this cell) of volume Ω_{cell}. Ω is the full crystal volume. $\Omega_{cell,\boldsymbol{n}}$ is the cell $\boldsymbol{n}$ with origin in $\boldsymbol{r}_{\boldsymbol{n}}$.

The integral factorization rule states that,[1]

$$\int_{\Omega} f(\boldsymbol{r})g(\boldsymbol{r})d^3r \cong \left(\int_{\Omega} f(\boldsymbol{r})d^3r\right)\left(\int_{\Omega_{cell}} g(\boldsymbol{r})\frac{d^3r}{\Omega_{cell}}\right) \tag{2.1}$$

To prove this, let us first accept that, for any function that varies slowly within the unit cells, as is the case of $f(\boldsymbol{r})$,

$$\sum_{\boldsymbol{n}} \Omega_{cell} f(\boldsymbol{r}_{\boldsymbol{n}}) \cong \int_{\Omega} f(\boldsymbol{r})d^3r \tag{2.2}$$

[1] The division by Ω_{cell} is necessary, among other reasons, to keep the dimensionality.

and then let us consider evident that for any function $h(\boldsymbol{r})$,

$$\int\limits_{\Omega_{cell,n}} h(\boldsymbol{r})d^3r = \int\limits_{\Omega_{cell,0}} h(\boldsymbol{r}+\boldsymbol{r}_n)d^3r \tag{2.3}$$

Taking this into account, we can write

$$\int\limits_{\Omega} f(\boldsymbol{r})g(\boldsymbol{r})d^3r = \sum_n \int\limits_{\Omega_{cell,n}} f(\boldsymbol{r})g(\boldsymbol{r})d^3r = \sum_n \int\limits_{\Omega_{cell,0}} f(\boldsymbol{r}+\boldsymbol{r}_n)g(\boldsymbol{r}+\boldsymbol{r}_n)d^3r \tag{2.4}$$

Defining $\Delta_n(\boldsymbol{r}) = f(\boldsymbol{r}+\boldsymbol{r}_n) - f(\boldsymbol{r}_n)$ and taking also into account the periodicity of $g(\boldsymbol{r})$,

$$\begin{aligned}\int\limits_{\Omega} f(\boldsymbol{r})g(\boldsymbol{r})d^3r &= \left(\int\limits_{\Omega_{cell,0}} g(\boldsymbol{r})\frac{d^3r}{\Omega_{cell}}\right)\sum_n f(\boldsymbol{r}_n)\Omega_{cell} + \sum_n \Omega_{cell}\int\limits_{\Omega_{cell,0}} \Delta_n(r)g(\boldsymbol{r})\frac{d^3r}{\Omega_{cell}} \\ &\cong \left(\int\limits_{\Omega} f(\boldsymbol{r})d^3r\right)\left(\int\limits_{\Omega_{cell,0}} g(\boldsymbol{r})\frac{d^3r}{\Omega_{cell}}\right) + \int\limits_{\Omega}\left(\int\limits_{\Omega_{cell,0}} \Delta_n(r)g(\boldsymbol{r})\frac{d^3r}{\Omega_{cell}}\right)d^3r\end{aligned} \tag{2.5}$$

Since our assumption is that the variation of $f(\boldsymbol{r}_n)$ in the $\Omega_{cell,0}$ cell is negligible and therefore $\Delta_n(r)$ is also negligible, the last term of the second line expression is negligible so that Eq. (2.1) is proven.

A second order approximation may help us to keep a clear idea of the approximation involved in using Eq. (2.1). Let us further consider the last term in Eq. (2.5). For this, let us develop $\Delta_n(r)$ in series around the origin to the first order,

$$\Delta_n(r) = f(\boldsymbol{r}+\boldsymbol{r}_n) - f(\boldsymbol{r}_n) \cong \nabla f(\boldsymbol{r}_n)\cdot\boldsymbol{r} \tag{2.6}$$

where $\nabla f(\boldsymbol{r}_n)$ are the gradients of $f(\boldsymbol{r}_n)$ calculated at specific points $\boldsymbol{r}_n$; this forms a set of constant vectors,

$$\sum_n \Omega_{cell}\int\limits_{\Omega_{cell,0}} \Delta_n(r)g(\boldsymbol{r})\frac{d^3r}{\Omega_{cell}} \cong \sum_n \Delta\Omega_{cell}\nabla f(\boldsymbol{r}_n)\cdot\left(\int\limits_{\Omega_{cell,0}} \boldsymbol{r}g(\boldsymbol{r})\frac{d^3r}{\Omega_{cell}}\right) \tag{2.7}$$

Assuming that $\nabla f(\boldsymbol{r}_n)$ varies slowly in each unit cells so allowing us to apply Eq. (2.2), we can write, in a second order approximation,

$$\int_{\Omega} f(\boldsymbol{r})g(\boldsymbol{r})d^3r \cong \left(\int_{\Omega} f(\boldsymbol{r})d^3r\right)\left(\int_{\Omega_{cell,0}} g(\boldsymbol{r})\frac{d^3r}{\Omega_{cell}}\right) + \left(\int_{\Omega} \nabla f(\boldsymbol{r_n})d^3r\right) \cdot \left(\int_{\Omega_{cell,0}} \boldsymbol{r}g(\boldsymbol{r})\frac{d^3r}{\Delta\Omega_{cell}}\right) \tag{2.8}$$

In this equation the last term is small mainly because r, when restricted to the unit cell at the origin, is small. For further discussion: it will be common that $\int_{\Omega} f(\boldsymbol{r})d^3r$ is normalized to one. If $f(\boldsymbol{r})$ is not zero in a radius R its height in this region will be of the order of $3/4\pi R^3$ (in a region comprising many unit cells), its derivative will be of the order of $3/4\pi R^4$ and $\int_{\Omega} \nabla f(\boldsymbol{r_n})d^3r$ will be of the order of $1/R$. On the other hand, it will be also common that $\int_{\Omega_{cell,0}} g(\boldsymbol{r})d^3r/\Omega_{cell}$ will also be one and $\int_{\Omega_{cell,0}} \boldsymbol{r}g(\boldsymbol{r})d^3r/\Omega_{cell} = r_{incell}$ with r_{incell} a value smaller than the unit cell average radius. In consequence the last term of Eq. (2.8) is of the order of r_{incell}/R, which is assumed to be small. Of course this is a rough calculation that has to be taken with care. For more accuracy, the value of the last term as compared with the first term (equal to one for the normalized functions) clarifies the accuracy of the approximation.

2.1.2 The Envelope Equation in a Nanostructured Semiconductor

Let $\Xi(\boldsymbol{r}, t)$ be a solution of the time-independent Schrödinger equation (TISE)

$$(H + U)\Xi(\boldsymbol{r}) = E\Xi(\boldsymbol{r}) \tag{2.9}$$

where

$$H = -(\hbar^2/2m_0)\nabla^2 + U_L(\boldsymbol{r}) \tag{2.10}$$

$U_L(\boldsymbol{r})$ being a periodic lattice potential and $U = U(\boldsymbol{r})$ a mesoscopically varying potential,[2] for instance, in the space charge region of a junction. It is well know that, in a homogeneous semiconductor, the eigenfunctions $H|\nu, \boldsymbol{k}\rangle = E_\nu(\boldsymbol{k})|\nu, \boldsymbol{k}\rangle$ are the Bloch functions. In $E_\nu(\boldsymbol{k})$, $\boldsymbol{k}$ can be considered any vector of the reciprocal space within the 1st Brillouin zone or, more often, any of the nodes of a lattice when the

[2] Note that, in the non-relativistic limit, the energies in the Schrödinger equations do not have a defined energy origin. However, the origin for E and U_L have to be the same and U must be considered as a potential energy that is added to U_L.

configuration space is limited to a finite, although large, volume Ω. Therefore, we may also write $E_v(\boldsymbol{k}) = E_{v,\boldsymbol{k}}$.

The *energy dispersion* $E_v(\boldsymbol{k})$ is a function known for many semiconductors, at least in the region of the $\boldsymbol{k}$-space where it is of interest. In many cases it is taken as parabolic. In this chapter,

$$E_v(\boldsymbol{k}) = E_{v,0} + \hbar^2 k^2/2m^* \tag{2.11}$$

with m^* being the effective mass and $E_{v,0}$ the extremum, or edge, of the band v, for instance the CB bottom. The effective mass m^* is approximately isotropic for zincblende semiconductors although it is direction-dependent for some other semiconductors. Note that Eq. (2.11) is a second order development of $E_v(\boldsymbol{k})$ and $E_v(0) = E_{v,0}$ is an extremum (maximum or minimum). In general, $E_v(\boldsymbol{k})$ can be more complex and may admit a development in series of (k_x, k_y, k_z) that, in any case, has to be symmetric with respect to the origin [2]. In nanostructured semiconductors, $E_{v,0}$ vary spatially with the nanostructure, forming an additional mesoscopic potential that, as we shall see, for all practical effects may be added to the mesoscopic potential U, the latter often being zero. It has to be born in mind that the Bloch function solutions are strictly valid only in homogeneous semiconductors; the validity of a dependence on $\boldsymbol{r}$ in $E_v(\boldsymbol{k})$ has to be taken with caution.

The Bloch functions are[3]

$$|v,\boldsymbol{k}\rangle \equiv u_{v,\boldsymbol{k}}(\boldsymbol{r})\frac{e^{i\boldsymbol{k}\cdot\boldsymbol{r}}}{\sqrt{\Omega}}; \quad \langle v',\boldsymbol{k}' \mid v,\boldsymbol{k}\rangle \equiv \int_{\Omega} u^*_{v',\boldsymbol{k}'}u_{v,\boldsymbol{k}}e^{i(\boldsymbol{k}-\boldsymbol{k}')\cdot\boldsymbol{r}}\frac{d^3\boldsymbol{r}}{\Omega} = \delta_{v,v'}\delta_{\boldsymbol{k},\boldsymbol{k}'} \tag{2.12}$$

($\equiv$ means equal by definition) with $u_{v,\boldsymbol{k}}(\boldsymbol{r})$ a function with the translational symmetry of the lattice. When $\boldsymbol{k} = \boldsymbol{k}'$, the orthonormality condition becomes

$$\int_{\Omega} u^*_{v',\boldsymbol{k}}u_{v,\boldsymbol{k}}\frac{d^3r}{\Omega} = \delta_{v,v'} = \frac{N_{cells}}{\Omega}\int_{\Omega_{cell}} u^*_{v',\boldsymbol{k}}u_{v,\boldsymbol{k}}d^3r = \int_{\Omega_{cell}} u^*_{v',\boldsymbol{k}}u_{v,\boldsymbol{k}}\frac{d^3r}{\Omega_{cell}} \tag{2.13}$$

where N_{cells} is the number of unit cells in the crystal. This gives an orthonormalization rule (similar to the one in Sect. 2.1.1) for the periodic part of the Bloch functions; the normalization of the $u_{v,\boldsymbol{k}}$ functions is obtained by average-normalizing them to one in a single unit cell. The orthogonality is always fulfilled.

We want to prove that the full wavefunction, $\Xi(\boldsymbol{r},t)$, can be approximately expressed as the product of a periodic function (the periodic part of a Bloch function), $u_{v,0}(\boldsymbol{r})$, and an *envolvent*, $\Phi(\boldsymbol{r},t)$, that is,

$$\Xi(\mathbf{r},t) \cong u_{v,0}(\mathbf{r})\Phi(\mathbf{r},t) \tag{2.14}$$

[3] Some authors, and more frequently those working in the k·p method, call Bloch functions to their periodic part, $u_{v,\boldsymbol{k}}(\boldsymbol{r})$.

where the *envolvent* fulfils the effective mass equation, defined as

$$(E_v(-i\nabla) + U)\Phi(\boldsymbol{r}, t) = E\Phi(\boldsymbol{r}, t) \tag{2.15}$$

($E_v(-i\nabla)$ is $E_v(\boldsymbol{k})$ where $\boldsymbol{k} \to -i\nabla$) under certain conditions to be discussed later. Strictly speaking, the transformation rule $E_v(-i\nabla) \to E_v(\boldsymbol{k})$ is a consequence of the Fourier transforms and is not straightforwardly valid if $E_v(\boldsymbol{k})$ also depends on $\boldsymbol{r}$. In this book we usually ignore this difficulty.

It will be proven later that, in the specific case of E($\boldsymbol{k}$) given by Eq. (2.11), the effective mass equation is

$$-\frac{\hbar^2}{2m^*}\nabla^2\Phi + (E_{v,0} + U)\Phi = E\Phi \tag{2.16}$$

Even in a homogeneous material, the periodic part of the Bloch functions is not easily known. There are approximations, and group theory provides some properties of these functions, but in most cases the used properties are the result of an indirect measurement. In some cases, in particular in the so called *ab initio* quantum calculations, they are theoretically obtained, but this implies a heavy calculation burden and, ultimately, the accuracy of these calculations is to be tested against measurements.

As already mentioned, $E_{v,0}$ depends on the position; it is different in the nanostructure than in the host semiconductor, leading to a mesoscopic quasi-potential, different in each band. The same can be said of the effective mass m^*; however, its variation makes uncertain the application of the rule $\boldsymbol{k} \to -i\nabla$ and may lead to the adoption of a non-Hermitical kinetic energy operator (see e.g. [1, p. 191]). We shall come later again to this point. In this work, we shall consider m^* constant throughout the whole nanostructured semiconductor. Since in the bound states the envelope functions are concentrated in the nanostructure, the m^* adopted is the one corresponding to the nanostructure, whose material becomes, in this way, the *semiconductor of reference*; however, the adoption of the host material m^* should be better for extended envelope functions.

Even neglecting the dependence of m^* on the position due to the nanostructure, Eq. (2.11) has a dependence on $\boldsymbol{k}$ but also, through $E_{v,0}$, on $\boldsymbol{r}$. This will be taken into account in the proof to follow.

2.1.3 Proof of the Effective Mass Equation for Obtaining the Envolvent

The proof goes along the following path: first, using the Bloch functions of the homogeneous semiconductors of reference as a basis, we obtain a given solution of the nanostructure. It is found that, when the periodic part of the Bloch function of a band does not sensibly vary with $\boldsymbol{k}$ in the range of small $\boldsymbol{k}$s, the wavefunction for this band becomes decoupled from the other bands. Then, coefficients of the Bloch function development of the nanostructure solution obtained in this case are found to be the same as the ones resulting from solving the effective mass equation with a development of plane waves. This straightforwardly provides a proof of Eq. (2.14).

Let us start by developing $\Xi(\boldsymbol{r})$ in the basis of Bloch functions of the semiconductor of reference.

$$\Xi(\boldsymbol{r}) = \sum_{v,\boldsymbol{k}} \varphi_{v,\boldsymbol{k}} |v, \boldsymbol{k}\rangle \tag{2.17}$$

The coefficients of the Bloch functions may be obtained from the TISE (Eq. (2.9)) through a set of linear equations in the variables $\varphi_{v,\boldsymbol{k}}$. For this, we use Eq. (2.17) and the dispersion function of Eq. (2.11),

$$\sum_{v,\boldsymbol{k}} E_{v,0}\phi_{v,\boldsymbol{k}}|v,\boldsymbol{k}\rangle + \sum_{v,\boldsymbol{k}} \frac{\hbar^2 k^2}{2m^*}\phi_{v,\boldsymbol{k}}|v,\boldsymbol{k}\rangle + \sum_{v,\boldsymbol{k}} U\phi_{v,\boldsymbol{k}}|v,\boldsymbol{k}\rangle = E\sum_{v,\boldsymbol{k}} \phi_{v,\boldsymbol{k}}|v,\boldsymbol{k}\rangle \tag{2.18}$$

After putting primes in the indices of Eq. (2.18), this equation is left-multiplied by $\langle v, \boldsymbol{k}|$; with some rearrangement of terms, we obtain.

$$\left(\frac{\hbar^2 k^2}{2m^*} - E\right)\phi_{v,\boldsymbol{k}} + \sum_{v',\boldsymbol{k}'} \langle v,\boldsymbol{k}|E_{v',0} + U|v',\boldsymbol{k}'\rangle \phi_{v',\boldsymbol{k}'} = 0 \tag{2.19}$$

$E_{v,0} + U$ depends on the space coordinates; however, they are contained in matrix elements that involve an integration whose result depends only on $(\boldsymbol{k},\boldsymbol{k}')$ couples of permitted points in the Brillouin zone. For each band and each permitted value of $\boldsymbol{k}$ an equation is formed and the solution of the system of equations in Eq. (2.19) allows the calculation of all the $\phi_{v,\boldsymbol{k}}$ values. Note that the procedure is rigorous and the band edge position dependence has been grouped with the slow varying potential U. A position dependence of the effective mass is more difficult to treat.

Let us analyze the matrix elements. For this we take into account that $E_{v,0} + U$ are slowly varying functions and we assume, for the moment, that the only matrix elements of interest are those of small $\boldsymbol{k}$, so that $\exp[i(\boldsymbol{k} - \boldsymbol{k}') \cdot \boldsymbol{r}]$ are slow-varying functions. The integral factorization rule of Eq. (2.1) can then be applied and we also assume that, at least for a certain range, $u_{b,\boldsymbol{k}}(\boldsymbol{r})$ is independent of $\boldsymbol{k}$ in this small range of interest and therefore equal to $u_{b,\boldsymbol{k}'}(\boldsymbol{r})$. Using also the orthonormalization rule in Eq. (2.13), we can write

$$\begin{aligned}\langle b,\boldsymbol{k}|E_{v',0} + U|v',\boldsymbol{k}'\rangle &\cong \int_{\Omega} (E_{v',0} + U)\mathrm{e}^{i(\boldsymbol{k}'-\boldsymbol{k})\cdot\boldsymbol{r}} \frac{d^3r}{\Omega} \int_{\Omega cell} u^*_{b,\boldsymbol{k}'} u_{v',\boldsymbol{k}'} \frac{d^3r}{\Omega_{cell}} \\ &= \langle \boldsymbol{k}'|E_{b,0} + U|\boldsymbol{k}\rangle \delta_{b,v'}\end{aligned} \tag{2.20}$$

where $|\boldsymbol{k}\rangle$ represents the plane wave

$$|\boldsymbol{k}\rangle \equiv \frac{e^{i\boldsymbol{k}\cdot\boldsymbol{r}}}{\sqrt{\Omega}} \tag{2.21}$$

Using Eq. (2.20), Eq. (2.19) will be written as

$$\left(\frac{\hbar^2 k^2}{2m^*} - E\right)\phi_{b,\boldsymbol{k}} + \sum_{\boldsymbol{k}'} \langle \boldsymbol{k}|E_{b,0} + U|\boldsymbol{k}'\rangle \phi_{b,\boldsymbol{k}'} = 0 \tag{2.22}$$

The condition for this to happen is that $u_{b,\boldsymbol{k}}(\boldsymbol{r})$ be independent of $\boldsymbol{k}$. Note that now this equation, corresponding to band b, is decoupled of all the other bands.

Let us now develop the effective mass Eq. (2.16) in plane waves

$$\Phi(\boldsymbol{r}) = \sum_{\boldsymbol{k}} \phi_{\boldsymbol{k}} |\boldsymbol{k}\rangle \tag{2.23}$$

(note that $|v, \boldsymbol{k}\rangle = u_{v,\boldsymbol{k}}(\boldsymbol{r})|\boldsymbol{k}\rangle$). This development is nothing more than the three-dimensional Fourier transform of $\Phi(\boldsymbol{r})$.

Substituting Eq. (2.23) into the effective mass Eq. (2.16) for the band b we obtain

$$\sum_{\boldsymbol{k}} \phi_{\boldsymbol{k}} \frac{\hbar^2 k^2}{2m^*} |\boldsymbol{k}\rangle + \sum_{\boldsymbol{k}} \phi_{\boldsymbol{k}} (E_{b,0} + U)|\boldsymbol{k}\rangle = E \sum_{\boldsymbol{k}} \phi_{\boldsymbol{k}} |\boldsymbol{k}\rangle \tag{2.24}$$

After putting primes on the indices of Eq. (2.24), this equation is left-multiplied by $\langle \boldsymbol{k}|$; with some rearrangement of terms, we obtain

$$\left(\frac{\hbar^2 k^2}{2m^*} - E\right)\phi_{\boldsymbol{k}} + \sum_{\boldsymbol{k}'} \langle \boldsymbol{k}|E_{b,0} + U|\boldsymbol{k}'\rangle \phi_{\boldsymbol{k}'} = 0 \tag{2.25}$$

This equation replicates Eq. (2.22) by taking $\phi_{\boldsymbol{k}} = \phi_{b,\boldsymbol{k}}$ for the given band b. Taking into account the plane wave development in Eq. (2.23), for a solution in the band b, Eq. (2.17) becomes,

$$\Xi(\boldsymbol{r}) = \sum_{\boldsymbol{k}} \phi_{\boldsymbol{k}} |\boldsymbol{k}\rangle u_{b,0}(\boldsymbol{r}) = \Phi(\boldsymbol{r}) u_{b,0}(\boldsymbol{r}) \tag{2.26}$$

So demonstrating Eq. (2.14) and proving that $\Phi(\boldsymbol{r})$ is a solution of the effective mass Eq. (2.16), with the parameters (effective mass and band-edge profile) of the band b.

Some of the assumptions may be tested now. Since $E_{v,0} + U$ varies slowly because of its big size compared to the lattice constant, for large values of $\boldsymbol{k} - \boldsymbol{k}'$,

$$\langle \boldsymbol{k}|U + E_{v,0}|\boldsymbol{k}'\rangle = \int_{\Omega} (E_{v,0} + U)\mathrm{e}^{i(\boldsymbol{k}-\boldsymbol{k}')\cdot\boldsymbol{r}} \frac{d^3\boldsymbol{r}}{\Omega} \cong 0 \tag{2.27}$$

This implies that, for large values of $\boldsymbol{k}$, $(E_{v,\boldsymbol{k}} - E)\phi_{\boldsymbol{k}}(t) \cong 0$ and $\phi_{\boldsymbol{k}}(t) \cong 0$. Actually, this is a well known property of the Fourier transform: extended $\boldsymbol{r}$-functions have a localized $\boldsymbol{k}$-Fourier transform. This confirms that the large values of $\boldsymbol{k}$ lack interest in the context of the problem examined here.

The constant character of $u_{v,\boldsymbol{k}}$ in the restricted domain of useful $\boldsymbol{k}$s is not to be taken for granted. In practice, the one-band approximation will be reasonable for the CB of most semiconductors with zincblende structure, but it is not a general property. For cases in which this condition cannot be assumed, multiple-band effective mass equations have been developed, as is studied later in this book.

2.1.4 Elements of Matrix for the Intraband Absorption

The IB → CB absorption, that for reasons to be seen later, is called intraband absorption, is proportional to the dipole [3, 4] element of matrix $|\langle \Xi'|\boldsymbol{\varepsilon}\cdot\boldsymbol{r}|\Xi\rangle|^2$, where the two wavefunctions describe two different states (at least one of them bound around the nanostructure). In this equation, $\boldsymbol{\varepsilon}$ is the polarization vector of the incident photon. The full expression for the absorption coefficient will be explained in Sect. 2.4.

Using Eq. (2.14) and applying the Integral Factorization Rule of Eq. (2.1) we can write

$$\begin{aligned} |\langle \Xi'|\boldsymbol{\varepsilon}\cdot\boldsymbol{r}|\Xi\rangle|^2 = \left|\langle u_{v,0}\Phi'|\boldsymbol{\varepsilon}\cdot\boldsymbol{r}|u_{v,0}\Phi\rangle\right|^2 &= \left|\int_{\Omega cell} u^*_{v,0}u_{v,0}\frac{d^3r}{\Omega_{cell}}\int_{\Omega}\Phi'^*\boldsymbol{\varepsilon}\cdot\boldsymbol{r}\Phi d^3r\right|^2 \\ &= \left|\int_{\Omega}\Phi'^*\boldsymbol{\varepsilon}\cdot\boldsymbol{r}\Phi d^3r\right|^2 \end{aligned} \tag{2.28}$$

Therefore, the intraband absorption calculations, when at least one wavefunctions is bound, can be performed by using just the envelope functions and disregarding the periodic part of the Bloch functions.

When the two states in the transition are extended along the whole semiconductor, the element of matrix used here is not valid [5]. Another element of matrix is then to be used, as explained in Chap. 5 of this book.

2.1.5 The Variable-Effective-Mass Envelope Hamiltonian

We assume Eq. (2.11) to be the position dependent dispersion equation for a nanostructured material, and use the Bloch functions of a certain reference homogeneous semiconductor as a basis.

As before, let us now develop $\Xi(\boldsymbol{r})$ in the basis of the Bloch functions of Eq. (2.17). The coefficients of the Bloch functions (corresponding to a discrete set of values of $\boldsymbol{k}$ in the first Brillouin zone) may be obtained from the TISE Eq. (2.9) through a set of linear equations in the variables $\phi_{v,k}$. We substitute Eqs. (2.11) and (2.17) in Eq. (2.9)

$$\sum_{v,k} \phi_{v,k} E_{v,0}(\boldsymbol{r})|v,\boldsymbol{k}\rangle + \sum_{v,k} \phi_{v,k} \frac{\hbar^2 k^2}{2m^*(\boldsymbol{r})}|v,\boldsymbol{k}\rangle + \sum_{v,k} \phi_{v,k} U(\boldsymbol{r})|v,\boldsymbol{k}\rangle = E \sum_{v,k} \phi_{v,k}|v,\boldsymbol{k}\rangle \tag{2.29}$$

and, multiplying on the left by $\langle b,\boldsymbol{k}|$ (after having put primes to all the running indices), we obtain the desired system of equations

$$-E\phi_{b,k} + \sum_{v',k'} \phi_{v',k'} \langle b,\boldsymbol{k}|E_{v',0}(\boldsymbol{r}) + \frac{\hbar^2 \boldsymbol{k}'^2}{2m^*(\boldsymbol{r})} + U(\boldsymbol{r})|v',\boldsymbol{k}'\rangle = 0 \tag{2.30}$$

All the dependence on $\boldsymbol{r}$ is in the second term sum. However, it is contained in matrix elements that involve an integration whose result depends only on $(\boldsymbol{k},\boldsymbol{k}')$ couples of permitted points in the Brillouin zone. For each band and each permitted value of $\boldsymbol{k}$, an equation is formed and the solution of the system of equations in Eq. (2.30) allows the calculation of all the $\phi_{v,k}$ values.

Notice that we have left-multiplied by the bra state $\langle b,\boldsymbol{k}|$ corresponding to the band in which $u_{b,k}(\boldsymbol{r})$ is independent of $\boldsymbol{k}$. Therefore, with the same arguments used in the discussion of Eq. (2.20), we can now write

$$\begin{aligned}
&\langle b,\boldsymbol{k}|E_{v',0}(\boldsymbol{r}) + \frac{\hbar^2 k'^2}{2m^*(\boldsymbol{r})} + U(\boldsymbol{r})|v',\boldsymbol{k}'\rangle \\
&\quad \cong \int_{\Omega} (E_{v',0}(\boldsymbol{r}) + \frac{\hbar^2 k'^2}{2m^*(\boldsymbol{r})} + U) e^{i(\boldsymbol{k}'-\boldsymbol{k})\cdot\boldsymbol{r}} \frac{d^3r}{\Omega} \int_{\Omega cell} u^*_{b,k'} u_{v',k'} \frac{d^3r}{\Omega_{cell}} \\
&\quad = \langle \boldsymbol{k}'|E_{b,0} + \frac{\hbar^2 k'^2}{2m^*(\boldsymbol{r})} + U|\boldsymbol{k}\rangle \delta_{b,v'}
\end{aligned} \tag{2.31}$$

So that the equation referring to band b,

$$-E\phi_{b,k} + \sum_{k'} \phi_{b,k'} \langle b,\boldsymbol{k}|E_{b,0}(\boldsymbol{r}) + U(\boldsymbol{r})|b,\boldsymbol{k}'\rangle + \sum_{k'} \phi_{b,k'} \langle b,\boldsymbol{k}|\frac{\hbar^2 k'^2}{2m^*(\boldsymbol{r})}|b,\boldsymbol{k}'\rangle = 0 \tag{2.32}$$

has been decoupled from all the terms belonging to other bands. The last sum we have separated from the preceding one will deserve our attention.

Let us look for a suitable effective mass equation. The application of the transformation rule $\boldsymbol{k} \to -i\nabla$ will lead to the following effective mass equation[4]

$$i\nabla\left(\frac{\hbar^2}{2m^*(\boldsymbol{r})}\right)i\nabla\Phi + \left(E_{b,0}(\boldsymbol{r}) + U(\boldsymbol{r})\right)\Phi = E\Phi \tag{2.33}$$

where, in the position-dependent term containing $\boldsymbol{k}^2$, this transformation has been made according to the rule $\boldsymbol{k}^2 f(\boldsymbol{r}) \to i\nabla f(\boldsymbol{r}) i\nabla$ for any real function $f(\boldsymbol{r})$ of $\boldsymbol{r}$. This is necessary because the operator resulting from a straightforward application of the rule $\boldsymbol{k} \to -i\nabla$ is not Hermitical (see for instance [3, Chap. VII] or [1, Chap. 6]).

Let us now develop this equation in basis of plane waves. By following the same steps used to get Eq. (2.25) we obtain

$$-E\phi_{\boldsymbol{b}}(E) + \sum_{\boldsymbol{k}'} \phi_{\boldsymbol{k}'}\langle \boldsymbol{b}|E_{b,0}(\boldsymbol{r}) + U(\boldsymbol{r})|\boldsymbol{k}'\rangle + \sum_{\boldsymbol{k}'} \phi_{\boldsymbol{k}'}\langle \boldsymbol{b}|\frac{\hbar^2 \boldsymbol{k}\boldsymbol{k}'}{2m^*(\boldsymbol{r})}|\boldsymbol{k}'\rangle = 0 \tag{2.34}$$

We can see that the last term of this equation differs somewhat from the last term of Eq. (2.32). This difference disappears if $m^*(\boldsymbol{r})$ is constant; in this case both equations become

$$-E\phi_{b,\boldsymbol{k}} + \sum_{\boldsymbol{k}'} \phi_{b,\boldsymbol{k}'}\langle b,\boldsymbol{k}|E_{b,0}(\boldsymbol{r}) + U(\boldsymbol{r})|b,\boldsymbol{k}'\rangle + \phi_{b,\boldsymbol{k}}\frac{\hbar^2 k^2}{2m^*} = 0 \tag{2.35}$$

which is the same as Eq. (2.22).

Thus we may conclude that the effective mass Eq. (2.33) sometimes proposed [6] to represent nanostructured semiconductors is not exactly the same as the one obtained when accepting the dispersion Eq. (2.11) when we accept that the effective mass is position dependent.

To avoid the conflict we use in this book a position independent effective mass in the way discussed before.

2.2 Box Shaped Quantum Dots

As described in the Introduction chapter, the confined levels associated with quantum dots (QDs) have been used to form a suitable IB. As represented in Fig. 2.1, large spacers have been used, of about 80 nm, between the layers of QDs,

---

[4] In this equation, the imaginary unit i is conserved attached to the gradient sign ∇. This is because $i\nabla$ is Hermitical. The advantage of using Hermitical operators is that $\langle\varphi|i\nabla|\psi\rangle = \int \varphi^* i\nabla\psi d^3r = \int (i\nabla\varphi)^*\psi d^3r$.

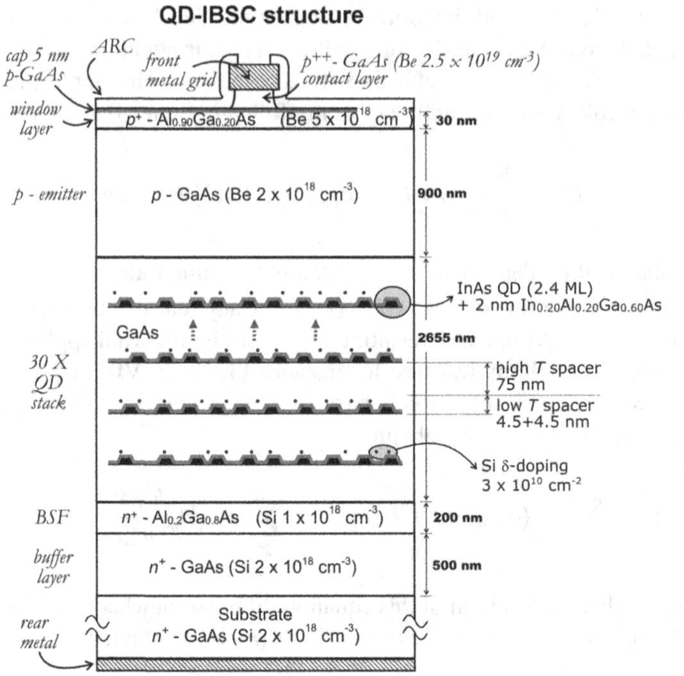

Fig. 2.1 Schematic of the solar cells used for the Isc-Voc measurements in Fig. 1.3. Reproduced with permission. © 2012, Elsevier [9]

instead of the more frequently used ones of about 20 nm, to avoid tunneling between QDs and permit the CB quasi-Fermi level to go beyond the IB levels [7] (only at low temperature) so permitting higher voltages. Indeed the big spacers decrease the volume density of QDs. These are actually only necessary in the edge of the QD region, where an electric field is produced to form the built-in potential. This involves six or seven QD layers near the p region. The use of field damping layers [8] is one proposed solution, and has been successfully used to avoid this drawback. However, it has not been used in the cell modeled in this chapter.

In particular, most of the calculations concerning the QD-IBSC along this chapter correspond to the sample SB of Ref. [7] (which is similar but not the same as that in Fig. 2.1). This is called the SOTA (state of the art) prototype in this book. Data for the calculations along the book are in Table 2.1.

2.2.1 Eigenvalues and Eigenfunctions in the Separation of Variables Approximation

The shape of the QDs is a matter of controversy and, as shown in this chapter, this shape has an important influence on their behavior. The most common understanding

Table 2.1 Parameters and dimensions used in the SOTA prototype

	Parameter	Symbol	Units	Value
QD (InAs)	Conduction band effective mass	m_{cb}/m_0		0.0294
	Light holes effective mass	m_{lh}/m_0		0.027
	Heavy holes effective mass	m_{hh}/m_0		0.333
	Split-off holes effective mass	m_{so}/m_0		0.076
	Conduction band offset	U_{cb}	eV	0.473
	Valence band offset for lh and hh	U_{vb}	eV	0.210
	Band offset for the split-off holes	U_{so}		0.341
	Index of refractions	n_{ref}		3.5
Barrier (GaAs)	Conduction band effective mass	m_{cb}/m_0		0.0613
	Barrier material bandgap (300 K)	E_g	eV	1.42
Dimensions	QD height	$2c$	nm	6
	QD base side	$2a = 2b$	nm	16
Array	Coverage factor of the QDs	F_s		0.1024
	QD layers density	N_l	cm^{-1}	125,000

of the InAs QD grown on (1,0,0) GaAs is that they are squat truncated pyramids of quadrangular base. Different authors have adopted different shapes in their calculations, with the probable motivation of fitting the shape to their models. In particular, spherical shape [11, 12], lens shape [13] or box shape [14, 15] have been selected. We also adopt the box shape, as seen in the lower part of Fig. 2.2, with dimensions $2a \times 2a \times 2c$.

The effective mass Eq. (2.16) can be written as

$$\frac{\hbar^2}{2m^*}\frac{\nabla^2 \Phi}{\Phi} + E = U(\boldsymbol{r}) \tag{2.36}$$

where the right-hand-side term is the band-edge position that that in the rest of this part will be denoted as $U(\boldsymbol{r})$ instead of $(E_{b,0} + U)$, despite the fact that $E_{b,0}(\boldsymbol{r})$, that in most cases will be $E_C(\boldsymbol{r})$, is the only non-null term of $(E_{b,0} + U)$. The energy origin is arbitrary and, for the calculations, the zero is set at the bottom of the potential well inside the QD (i.e. at the dot material conduction-band edge) and V outside it. In symbolic language ($\forall$ = for all, $\wedge$ = and, $\vee$ = or),

$$U(\boldsymbol{r}) = \begin{cases} 0 & \forall |x| < a \wedge |y| < a \wedge |z| < c \\ V & \forall |x| \geq a \vee |y| \geq a \vee |z| \geq c \end{cases} \tag{2.37}$$

However, for presentation purposes, the zero shall be set at the barrier material conduction band edge, which means that V must be subtracted from all the energy results.

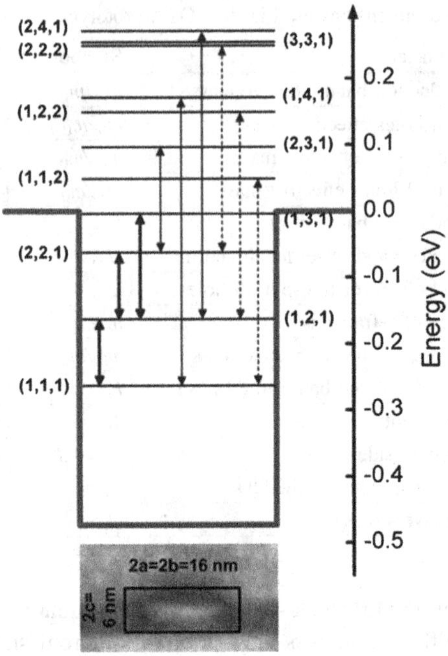

Fig. 2.2 Potential well and separation-of-variables energy levels in a box shaped potential well with the dimensions shown in the figure. *Solid arrows* are permitted transitions under vertical or isotropic illumination; *dotted* are permitted under isotropic illumination only. *Thicker lines* show the main electron flow. Levels (121) collect the electrons received directly from the VB and via the levels (111) and (221) (the latter going downwards) and send them to the (131) level. *Below* TEM photograph of a QD showing its dimensions. Reproduced with permission. © 2011, the authors [10]

2.2.1.1 One-Dimensional Solutions

The functions $\xi(x)$, $\psi(y)$, $\zeta(z)$ are defined by

$$\frac{\hbar^2}{2m^*}\frac{d^2\xi/dx^2}{\xi} + E_x = \begin{cases} 0 & \forall\,|x|<a \\ V & \forall\,|x|\geq a \end{cases} \tag{2.38}$$

and similarly for $\psi(y)$, $\zeta(z)$ (in the latter case using c instead of a as the well boundary).

Finding the solutions for $\xi(x)$ (or for $\psi(y)$, $\zeta(z)$) constitutes a simple exercise of differential equations. In this context, a discussion is provided e.g. in [16]. For $E < V$ (the subindex x is dropped in this subsection) bounded solutions, different from the trivial $\xi(x) \equiv 0$, are even ($\cos(kx)$) or odd ($\sin(kx)$) harmonic functions inside the well, flanked by fading exponential functions outside it ($\exp(-\kappa x)\ \forall x \geq a$). The wavefunctions must be normalizable in an infinite space, and, as such, non-fading exponential functions outside the QD are unphysical. The energy E, fading coefficient κ and k-values are related by

$$E = \hbar^2 k^2/2m^* = V - \hbar^2 \kappa^2/2m^* \tag{2.39}$$

The condition for cancellation of the non fading solutions is that the logarithmic derivative of the harmonic solution inside the QW must be matched to the logarithmic derivative of the fading solution outside it, that is,

$$\cot(k_n a) = k_n/\kappa_x = k_n a \Big/ \sqrt{\varpi^2 a^2 - k_n^2 a^2} \quad \text{for even functions}$$

$$-\tan(k_n a) = k_n/\kappa_x = k_n a \Big/ \sqrt{\varpi^2 a^2 - k_n^2 a^2} \quad \text{for odd functions} \tag{2.40}$$

$$\sqrt{\frac{2m^* V}{\hbar^2}} \equiv \varpi$$

The functions to the right and the left of the equalities above are represented in Fig. 2.3. The index n—a quantum number (QN)—denotes the different permitted energies in increasing order. Odd QNs correspond to even functions and vice versa.

Note that the rightmost side expression has a vertical asymptote at $ka = \varpi a$. There are positive branches of the minus-tangent and cotangent functions every $\pi/2$. Therefore, the number of confined states is the number of times $\pi/2$ is contained in ϖa plus one. The only parameter in the equations above is ϖa, or, in other words, $m^* V a^2$. This means that the number of confined states depends strongly on the size of the QD and, to a lesser extent, on the confining potential V and on the effective mass. The dependence on V and m^* is of the same importance, such that doubling the effective mass and halving the potential leaves the root unchanged, although this will also have the effect of halving the energies of the levels, as seen in Eq. (2.39). However, increasing the size while keeping the effective mass and potential unchanged increases greatly the number of confined states, as is visible by comparing plots (a) and (b) of Fig. 2.3.

Table 2.2 presents the values of k for the different QNs and the one-dimensional energy. The CB offset and QD dimensions are those in [17] and are derived from

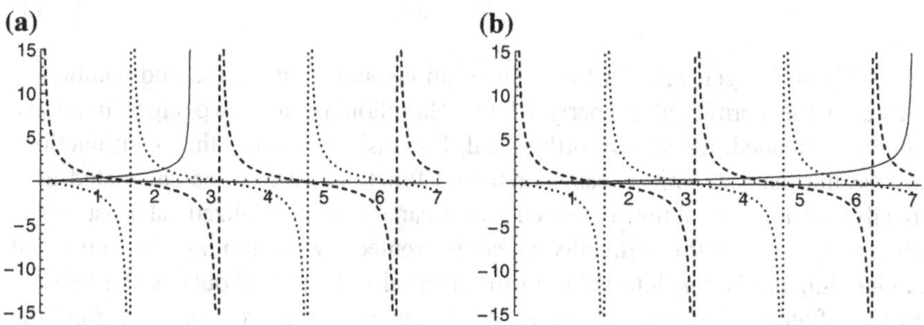

Fig. 2.3 The two members of the equations $\cot(k_n a) = k_n a/\sqrt{\varpi^2 a^2 - k_n^2 a^2}$ and $-\tan(k_n a) = k_n a/\sqrt{\varpi^2 a^2 - k_n^2 a^2}$ versus $k_n a$ for U = 0.473 eV, m = 0.0294 times the electron mass; **a** a = 3nm and **b** b = 8 nm. *Dashed curve* (cot) for even and *dotted* (−tan) for odd states. Second term of the equality, *solid line*

Table 2.2 Values of k (multiplied by the potential well half-width) and one-dimensional energy for the CB offset and QD dimensions in [17]. Energies are with respect to the QD material CB bottom

x-,*y*-eigenfunctions					*z*-eigenfunctions	
n	1	2	3	4	1	2
$k_x a$, $k_z c$	1.299	2.579	3.806	4.810	0.992	1.775
E_0 (eV)	0.034	0.135	0.292	0.469	0.142	0.453

$U = 0.473$ eV; $m^* = 0.0294\ m_0$; $a = b = 8$ nm; $c = 3$ nm

the data of the SOTA prototype (different of those in Fig. 2.3). We have used the InAs effective mass because the bound functions mainly span this material.

For $E \geq V$, the solution is harmonic with wavenumber k inside the potential well and also harmonic, even or odd, outside it but with a different value of the wavenumber k_e and a phase term. That is, they are of the form $\cos(k_e x - \theta)$ or $\sin(k_e x - \theta)$. Details can be found, e.g. in [17]. In this case,

$$E = V + \hbar^2 k_e^2 / 2m^* = \hbar^2 k^2 / 2m^* \tag{2.41}$$

and

$$\begin{aligned} (k_e/k)\cot(ka) &= \cot(k_e a - \theta) \\ (k_e/k)\tan(ka) &= \tan(k_e a - \theta) \end{aligned} \tag{2.42}$$

respectively for the even and odd functions.

For $E \geq V$, k_e can take any value and therefore it leads to a continuum spectrum of energies. Since the mathematics of continuum spectra is rather complicated, it is common to assume that the wavefunctions are restricted to a large but finite region (a segment of length $2L$, with large L, for one-dimensional cases, or a big parallelepiped for three-dimensional ones) and assume periodic conditions there. This leads [17] to

$$k_e L - \theta = \tilde{n}\pi/2 \tag{2.43}$$

where $\tilde{n}$ is an integer, odd for the even solutions and even for the odd solutions.

Due to the hermitical property of the Hamiltonian, any two eigenfunctions, bound or extended, are strictly orthogonal. For easier handling, the eigenfunctions must be normalized. This is easy for bound functions (it may be calculated analytically). Strictly speaking, extended states cannot be normalized (at least in the ordinary sense) but this difficulty is circumvented by integrating the square of the eigenfunction's absolute value in the interval ($-L$, $+L$). If outside the QD the harmonic function has amplitude one, this integration is approximately L (and the wavefunction norm is $L^{1/2}$) if L is much larger than the QD dimensions. L it is arbitrarily chosen but any measurable magnitude has, under this model, an expression containing L that cancels out the L dependence.

2.2.1.2 Three-Dimensional Solutions

By adding the one-dimensional Eq. (2.38) corresponding to the three coordinates we obtain

$$\frac{\hbar^2}{2m^*}\frac{\nabla^2\Phi}{\Phi} + E_{0,n_x,n_y,n_z} = U_0(\boldsymbol{r}) \tag{2.44}$$

where

$$\begin{aligned}\Phi(x,y,z) &= \xi(x)\psi(y)\zeta(z)\\ E_{0,n_x,n_y,n_z} &= E_{x,n_x} + E_{y,n_y} + E_{z,n_z}\end{aligned} \tag{2.45}$$

However, $U_0(\boldsymbol{r})$ is different from $U(\boldsymbol{r})$. As explained in [16, 18], it is the same inside the QD and outside it in front of the faces, but it takes the value $2V$ in front of the edges and of $3V$ in front of the corners. In symbolic language,

$$U_0(\boldsymbol{r}) = \begin{cases} 0 & \forall\,|x|<a \wedge |y|<a \wedge |z|<c\\ V & \forall\,|x|\geq a \wedge |y|<a \wedge |z|<c\\ V & \forall\,|x|<a \wedge |y|\geq a \wedge |z|<c\\ V & \forall\,|x|<a \wedge |y|<a \wedge |z|\geq c\\ 2V & \forall\,|x|\geq a \wedge |y|\geq a \wedge |z|<c\\ 2V & \forall\,|x|\geq a \wedge |y|<a \wedge |z|\geq c\\ 2V & \forall\,|x|<a \wedge |y|\geq a \wedge |z|\geq c\\ 3V & \forall\,|x|\geq a \wedge |y|\geq a \wedge |z|\geq c \end{cases} \tag{2.46}$$

Thus, the Hamiltonian we want to solve can be written as

$$H = -\frac{\hbar^2}{2m^*}\nabla^2 + U_0(\boldsymbol{r}) + U'(\boldsymbol{r}) = H_0 + U'(\boldsymbol{r}) \tag{2.47}$$

where the perturbation potential is

$$U'(\boldsymbol{r}) = U(\boldsymbol{r}) - U_0(\boldsymbol{r}) = \begin{cases} 0 & \forall\,|x|<a \wedge |y|<a \wedge |z|<c\\ 0 & \forall\,|x|\geq a \wedge |y|<a \wedge |z|<c\\ 0 & \forall\,|x|<a \wedge |y|\geq a \wedge |z|<c\\ 0 & \forall\,|x|<a \wedge |y|<a \wedge |z|\geq c\\ -V & \forall\,|x|\geq a \wedge |y|\geq a \wedge |z|<c\\ -V & \forall\,|x|\geq a \wedge |y|<a \wedge |z|\geq c\\ -V & \forall\,|x|<a \wedge |y|\geq a \wedge |z|\geq c\\ -2V & \forall\,|x|\geq a \wedge |y|\geq a \wedge |z|\geq c \end{cases} \tag{2.48}$$

(notice that here we have defined H as the effective mass Hamiltonian for the envolvent equation; in Sect. 2.1, H represented the lattice one-electron Hamiltonian). This perturbation is spatially limited to the corners and edges of the QD and is zero in the rest of the space. The separation-of-variables model assumes it is negligible. The separation-of-variables eigenfunctions can be labeled through three QNs: $|n_x, n_y, n_z\rangle = \xi_{n_x}\psi_{n_y}\zeta_{n_z} = \Phi_{n_x,n_y,n_z}$. They are the eigenfunctions of the Hamiltonian H_0. Each one-dimensional eigenfunction may be bound or extended. To distinguish this fact we can write $|\hat{n}_x, \tilde{n}_y, \hat{n}_z\rangle$ where $\hat{n}_x$ indicates that the x-eigenfunction is bound and $\tilde{n}_y$ denotes that the y-eigenfunction is extended. The H_0 eigenvalues are in Eq. (2.45) whose components may be found in Eqs. (2.39) or (2.41). As said before, we usually subtract V from the values obtained to refer the energies to the barrier material CB bottom.

While in one dimension, the bound states (BSs) energy was always below the potential well rim, this is not true in three-dimensional BSs, as shown in Fig. 2.2. These states have been considered by several researchers [19, 20], and sometimes are interpreted as resonances between plane waves; often they are called virtual bound states (VBSs). Their interpretation in the effective mass Hamiltonian equation is straightforward.

The set $|n_x, n_y, n_z\rangle$ (this notation is equivalent to (n_x, n_y, n_z)) forms an orthonormal basis that, nevertheless, is somewhat involved. It is formed of purely bound states $|\hat{n}_x, \hat{n}_y, \hat{n}_z\rangle$ (0E states), of purely extended states $|\tilde{n}_x, \tilde{n}_y, \tilde{n}_z\rangle$ (3E states) and of states that are mixed, with a single extended state, of the type $|\hat{n}_x, \hat{n}_y, \tilde{n}_z\rangle$ (and all the circular permutations: 1Ez, 1Ex, 1Ey states) and with a two extended states $|\tilde{n}_x, \tilde{n}_y, \hat{n}_z\rangle$ (and all the circular permutations: 2Exy, 2Eyz, 2Ezx states). The QN corresponding to the extended states is linked, as indicated in Eq. (2.43) to a certain value of L and it may be physically more meaningful to use k_e and the parity as the wavefunction definer, although, for calculations, we shall use the corresponding QN (the closest one of the same parity, once L is given). The relationship between the description based on QNs and k_e and parity are given by Eqs. (2.43) and (2.42).

Examples of H_0 eigenfunctions of the 0E, 1E and 2E sets are given in Fig. 2.4.

If H_0 is developed in this basis, it forms a diagonal matrix, although the ordering of the diagonal elements in this matrix is complicated. For instance we can start by

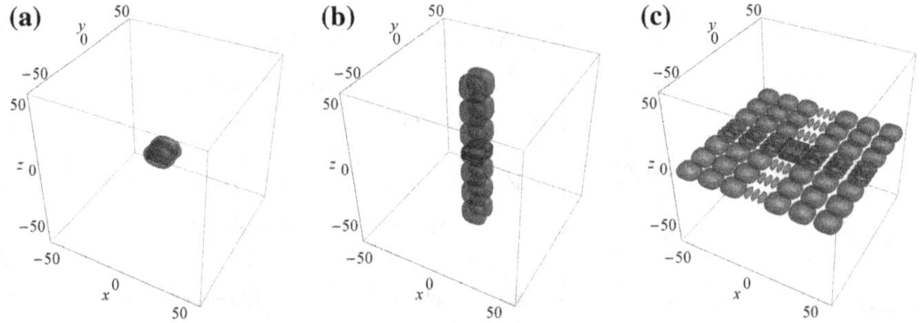

Fig. 2.4 Electron probability density contours. **a** Bound 0E (1,2,1) wavefunction; **b** filamentary 1Ez (1,2,8) wavefunction; **c** sheet-shaped 2Exy (5,8,1) wavefunction. Space units are in nm; the containing cube length is $2L = 120$ nm. SOTA prototype; the parallelepipedic box (in *green*, online) represents the QD, of $16 \times 16 \times 6$ nm^3. The contour drawn corresponds to 0.001 times the maximum probability density of the (1,1,1) 0E state. Reproduced with permission. © 2013, Elsevier [18]

ordering the finite number of states of the bound 0E states and then put the 3E states, but they are infinite, and we still have to situate the 2E and 1E states.

Note that, for extended states, the electron tends to be outside the QD. It might be reasonable that we use the GaAs effective mass for these states. However, this might compromise the orthogonality of these states with the bound states. Maybe in a problem not dealing with bound states at all it might be better to use the GaAs effective mass. Another option is to use the so called Bastard boundary conditions to be examined next.

2.2.2 Separation of Variables with Position-Dependent Effective Mass

This case has been discussed in Sect. 2.1.5 and it has been rejected for the mainstream of this book. Nevertheless the discussion is completed in this subsection.

The effective mass equation is

$$i\nabla\left(\frac{\hbar^2}{2m^*(\boldsymbol{r})}\right)i\nabla\Phi+\left(E_{b,0}(\boldsymbol{r})+U(\boldsymbol{r})\right)\Phi=E\Phi \tag{2.49}$$

For the region inside the QD the effective mass is constant and the variable effective mass Eq. (2.33) is identical to the one for constant effective mass Eq. (2.16); therefore the solutions are the same. The same can be said of the regions outside the QD and in front of the QD box faces, although, in this case, the effective mass is that of the barrier material and not that of the QD. The difference arrives at the interfaces. There, the one dimensional equation is

$$-\frac{d}{dx}\left(\frac{\hbar^2}{2m^*(x)}\right)\frac{d\xi(x)}{dx}+U(x)\xi(x)=E_x\xi(x) \tag{2.50}$$

By integrating this equation at the two sides of one interface,

$$\begin{gathered}-\left(\frac{\hbar^2}{2m^*(x)}\right)\frac{d\xi(x)}{dx}\bigg|_{a-\varepsilon}^{a+\varepsilon}=\int\limits_{a-\varepsilon}^{a+\varepsilon}(E_x-U(x))\xi(x)dx\cong 0\\ \text{or}\\ \left(\frac{\hbar^2}{2m^*_{GaAs}}\right)\frac{d\xi(a+\varepsilon)}{dx}=\left(\frac{\hbar^2}{2m^*_{InAs}}\right)\frac{d\xi(a-\varepsilon)}{dx}\end{gathered} \tag{2.51}$$

In summary, the quantity to be matched at the interface is the derivative of the wavefunction divided by the effective mass. This conditions is often called Bastard

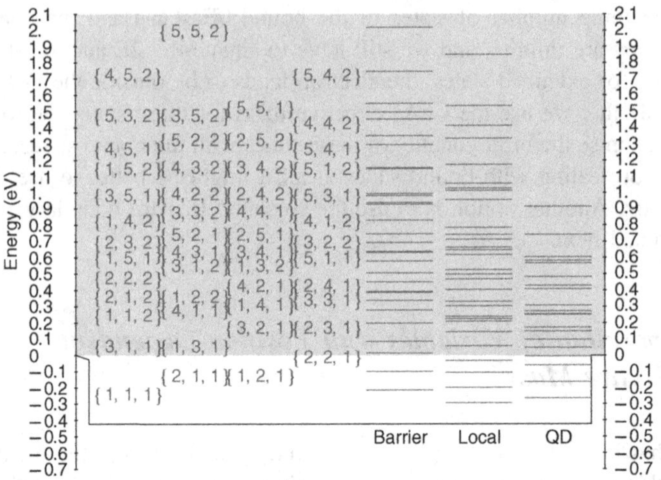

Fig. 2.5 Energy spectrum of the box shaped QD of the SOTA cell, using the barrier material effective mass (label barrier), the variable effective mass case (label local) and the QD effective mass (label QD). The QNs refer only to the barrier case

boundary condition [6]. It differs to the case of a spatially constant effective mass, for which just the derivative of the wavefunction must be matched.

Figure 2.5 presents the position of the fully confined energy levels for the SOTA cell when taking the barrier material effective mass (GaAs, $0.0613m_0$), the QD effective mass (InAs, $0.0294m_0$) and the variation contemplated in this subsection. It is interesting to observe the big difference between selecting the barrier material effective mass, which we estimated best for the study of extended wavefunctions, and the QD effective mass, which we considered best for confined states. The variable-effective-mass solution resembles the latter case for low energy and the former for high energy states. It must be pointed out that the variable effective mass presents a lower energy in the fundamental state than the case with the QD effective mass.

As we have discussed in Sect. 2.1.5, the variable-effective-mass approach does not clearly justify eigenfunctions which are the product of an envolvent times the periodic part of a Bloch function. Nonetheless, if we accept the position-variable dispersion Eq. (2.11), their spectrum is close to the two spectra we recommend for bound and extended functions, with the QD and barrier effective mass respectively. Given the Hermitical nature of the operator in Eq. (2.50), its eigenvalues are orthogonal. Therefore, the variable-effective-mass wavefunctions are a good practical choice for many problems. Nevertheless, we do not use them further in this book.

It is clear that also in this case the separation of variables, treated above, is only an approximation. In the edge and corner regions the solutions found do not fulfill the differential Eq. (2.33), for reasons associated to the variable effective mass, which are additional to those already discussed in Sect. 2.2.1.

2.3 Exact Resolution of the Mass Effective Equation with a Potential Box-Shaped Well

As discussed in Sect. 2.2.1.1, the separation-of-variables method produces an inexact solution to the differential Eq. (2.33). In this section, Eq. (2.33) is solved exactly, without using a separation of variables. The resulting exact Hamiltonian is going to be represented as a matrix in a basis of separation-of-variables wavefunctions, here called the standard basis, and then diagonalized. It closely follows Ref. [18], whose reading is recommended for further understanding.

2.3.1 *Integration in Regions*

As derived from the perturbation potential in Eq. (2.48), the matrix elements of the perturbation Hamiltonian $\langle n_{x,1}, n_{y,1}, n_{z,1} | U'(\boldsymbol{r}) | n_{x,2}, n_{y,2}, n_{z,2} \rangle$ require the integrations of wavefunctions restricted to a certain region of the space, multiplied by a multiple of V. This section is devoted to the calculation of these integrals.

The complex conjugation necessary in quantum mechanics for the internal products and matrix elements is not used here because all the functions are real.

In Tables 2.3 and 2.4, the values of the integral inside and outside the QD are presented for bound one-dimensional eigenfunctions in the x and z coordinates respectively. The column to the left represents the internal product; for different QNs it must be zero. The very small values observed in some cases are to be considered the noise background of our calculations. For small QNs, the probability of finding the electron outside the QD (look at the cases with equal QNs) is very small, so justifying the choice of the QD effective mass. For large QNs it may be more spread out. When the parity of the QNs is different, the product of wavefunctions is odd and its integral is zero inside the QD. Outside it, it takes different signs for positive and negative abscissas. When the QNs are different, but of the same parity, the integrals inside and outside the QD balance out.

For extended one-dimensional eigenfunctions of different QNs, the orthogonality is theoretically required and well verified in our calculations. Thus, for $L \rightarrow \infty$ (the amplitude of $\xi_{\tilde{n}_x}$ is rendered zero by the normalization),

$$\begin{aligned} &\int_{-a}^{a} \xi_{\tilde{n}_x} \xi_{\tilde{n}'_x} dx = 0 \\ &\int_{-\infty}^{-a} \xi_{\tilde{n}_x} \xi_{\tilde{n}'_x} dx + \int_{a}^{\infty} \xi_{\tilde{n}_x} \xi_{\tilde{n}'_x} dx = \delta_{\tilde{n}_x, \tilde{n}'_x} \end{aligned} \tag{2.52}$$

and for bound/extended eigenfunctions

Table 2.3 Integration in regions of bound one-dimensional eigenfunctions corresponding to the coordinates x or y

QNs	$\int_{-\infty}^{\infty} \xi_{\hat{n}_x}\xi_{\hat{n}'_x}dx$	$\int_{-\infty}^{a} \xi_{\hat{n}_x}\xi_{\hat{n}'_x}dx$	$\int_{-a}^{a} \xi_{\hat{n}_x}\xi_{\hat{n}'_x}dx$	$\int_{a}^{\infty} \xi_{\hat{n}_x}\xi_{\hat{n}'_x}dx$
1,1	1	0.00638387	0.987232	0.00638387
1,2	0	0.0133362	0	−0.0133362
1,3	1.606×10^{-16}	−0.0217575	0.0435151	−0.0217575
1,4	0	−0.0267906	0	0.0267906
2,2	1	0.0279778	0.944044	0.0279778
2,3	0	−0.0461097	0	0.0461097
2,4	-3.47531×10^{-17}	−0.059098	0.118196	−0.059098
3,3	1	0.0779097	0.844181	0.0779097
3,4	0	0.111237	0	−0.111237
4,4	1	0.336536	0.326929	0.336536

Table 2.4 Integration in regions of bound one-dimensional eigenfunctions corresponding to the coordinate z

QNs	$\int_{-\infty}^{\infty} \zeta_{\hat{n}_z}\zeta_{\hat{n}'_z}dz$	$\int_{-\infty}^{c} \zeta_{\hat{n}_z}\zeta_{\hat{n}'_z}dz$	$\int_{-c}^{c} \zeta_{\hat{n}_z}\zeta_{\hat{n}'_z}dz$	$\int_{c}^{\infty} \zeta_{\hat{n}_z}\zeta_{\hat{n}'_z}dz$
1,1	1	0.000375021	0.99925	0.000375021
1,2	0	0.00328518	0	−0.00328518
2,2	1	0.0345654	0.930869	0.0345654

$$\begin{gathered}\int_{-a}^{a} \xi_{\hat{n}_x}\xi_{\tilde{n}_x}dx = 0 \\ \int_{-\infty}^{-a} \xi_{\hat{n}_x}\xi_{\tilde{n}_x}dx + \int_{a}^{\infty} \xi_{\hat{n}_x}\xi_{\tilde{n}_x}dx = 0\end{gathered} \tag{2.53}$$

The y- and z-eigenfunctions behave similarly.

2.3.2 *The Exact Hamiltonian Matrix Elements, Eigenvalues and Eigenvectors*

The exact Hamiltonian matrix, when developed in eigenfunctions of the Hamiltonian H_0 of Eq. (2.27), is the sum of the matrix representation of H_0, which is a diagonal matrix, and the matrix representation of U', described in Eq. (2.48).

The matrix elements of the latter are

$$
\begin{aligned}
&\langle n_x, n_y, n_z | U' | n'_x, n'_y, n'_z \rangle \\
&= -V\Bigg[\int_{-a}^{a} \xi_{n_x}\xi_{n'_x}dx \left(\int_{-\infty}^{-a} \psi_{n_y}\psi_{n'_y}dy + \int_{a}^{\infty} \psi_{n_y}\psi_{n'_y}dy\right)\left(\int_{-\infty}^{-c} \zeta_{n_z}\zeta_{n'_z}dz + \int_{c}^{\infty} \zeta_{n_z}\zeta_{n'_z}dz\right) \\
&\quad + \left(\int_{-\infty}^{-a} \xi_{n_x}\xi_{n'_x}dx + \int_{a}^{\infty} \xi_{n_x}\xi_{n'_x}dx\right)\int_{-a}^{a} \psi_{n_y}\psi_{n'_y}dy \left(\int_{-\infty}^{-c} \zeta_{n_z}\zeta_{n'_z}dz + \int_{c}^{\infty} \zeta_{n_z}\zeta_{n'_z}dz\right) \\
&\quad + \left(\int_{-\infty}^{-a} \xi_{n_x}\xi_{n'_x}dx + \int_{a}^{\infty} \xi_{n_x}\xi_{n'_x}dx\right)\left(\int_{-\infty}^{-a} \psi_{n_y}\psi_{n'_y}dy + \int_{a}^{\infty} \psi_{n_y}\psi_{n'_y}dy\right)\int_{-c}^{c} \zeta_{n_z}\zeta_{n'_z}dz\Bigg] \\
&\quad - 2V\Bigg[\left(\int_{-\infty}^{-a} \xi_{n_x}\xi_{n'_x}dx + \int_{a}^{\infty} \xi_{n_x}\xi_{n'_x}dx\right)\left(\int_{-\infty}^{-a} \psi_{n_y}\psi_{n'_y}dy + \int_{a}^{\infty} \psi_{n_y}\psi_{n'_y}dy\right)\left(\int_{-\infty}^{-c} \zeta_{n_z}\zeta_{n'_z}dz + \int_{c}^{\infty} \zeta_{n_z}\zeta_{n'_z}dz\right)\Bigg]
\end{aligned}
\tag{2.54}
$$

where no distinction has been made between bound and extended states.

According to Eq. (2.53), the matrix elements linking a bound and an extended wavefunction in a given coordinate are zero. Therefore, the only non-zero matrix elements are those belonging to the same set (0E, 3E etc.) of those enumerated in Sect. 2.2.1.2. This is the same as saying that the Hamiltonian matrix is the direct sum of the Hamiltonian matrices in these sets. In other words, it is a diagonal of blocs, although many of these blocs are of infinite dimension.

2.3.2.1 Purely Bound 0E States

For the sizes and potential of the SOTA cell there are 4 QNs for the x and y one-dimensional eigenfunctions and 2 for the z one-dimensional eigenfunction. Thus, there are 32 H_0 eigenfunctions for BSs and VBSs. The eigenvalues are calculated as a sum of the one-dimensional eigenvalues, as shown in Eq. (2.45) using Table 2.2 where these eigenvalues are given. They are represented in Fig. 2.6a. The energy first-order approximation (see for instance [3])

$$
E_{1,n_x,n_y,n_z} = E_{0,n_x,n_y,n_z} + \langle n_x, n_y, n_z | U' | n_x, n_y, n_z \rangle \tag{2.55}
$$

is also presented there.

Notice that the first order approximation moves all the states downward. This shift is very little for the low energies but substantial for the high energies.

An exact calculation of the perturbed Hamiltonian, that is, the exact one within the model limitations, can be obtained by calculating the matrix

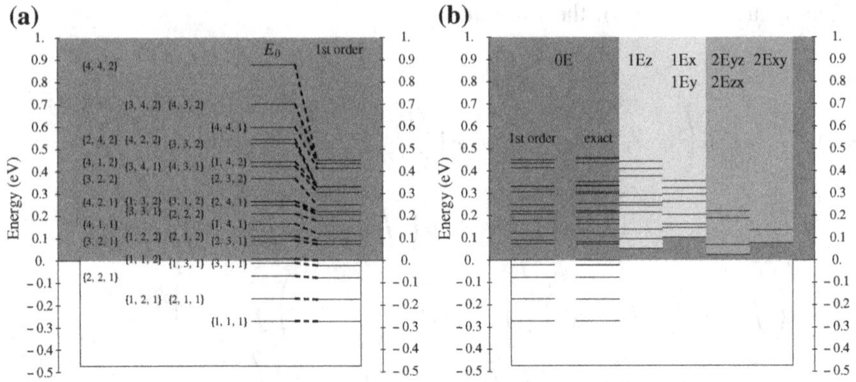

Fig. 2.6 **a** QD well and energy eigenvalues of $E_{0,nx,ny,nz}$ and their correction to the 1st order. The QNs corresponding to each energy line are also given. **b** Energy eigenvalues for the 1st order approximation, the same as in part (**a**), and from the exact calculations for the different sets of eigenfunctions. *Dark grey* corresponds to the energies of the continuum of 3E states. *Lighter* and *medium grey* correspond to the 1E and 2E states respectively, the *horizontal lines* show where the continuum spectrum starts above the energy of the bound part of the wavefunctions. One continuum of energies starts above each *horizontal line* in these two groups of sets. Reproduced with permission. © 2013, Elsevier [18]

$$\left\langle n_x, n_y, n_z \middle| H \middle| n'_x, n'_y, n'_z \right\rangle = \left\langle n_x, n_y, n_z \middle| H_0 \middle| n'_x, n'_y, n'_z \right\rangle + \left\langle n_x, n_y, n_z \middle| U' \middle| n'_x, n'_y, n'_z \right\rangle \tag{2.56}$$

and obtaining the eigenvalues and eigenvectors. Of course, H_0, developed in the standard basis, is a diagonal matrix of elements equal to E_{0,n_x,n_y,n_z}. The eigenvalues are shown in Fig. 2.6b side by side with the 1st order approximations (which are also in part (a) of the figure) for easy comparison. Changes are small but visible for the high energies. It must be stressed that, for the exact calculations, the energy levels cannot be related to any set of QNs. In fact, the eigenstates are linear combinations of the unperturbed Hamiltonian eigenstates that are the product of one-dimensional wavefunctions, each with one QN. This relationship is represented in Fig. 2.7. The absolute value of the coefficient of $\left| n_x, n_y, n_z \right\rangle$ is represented in a gray scale (1 black; 0 white). The numeric values are given in the Supplemental Material of Ref. [18].

It can be seen that, in general, there are a small number of dominant unperturbed states corresponding to each exact state. Diagonal blocks of four boxes are often found sometimes corresponding to double degenerate states. The order of the unperturbed eigenstates on the top follows the order of the increasing E_{0,n_x,n_y,n_z} energies.

As an example, the bound states within the bandgap for the case in study (retaining only the terms producing a shadow in Fig. 2.7) are presented in Table 2.5.

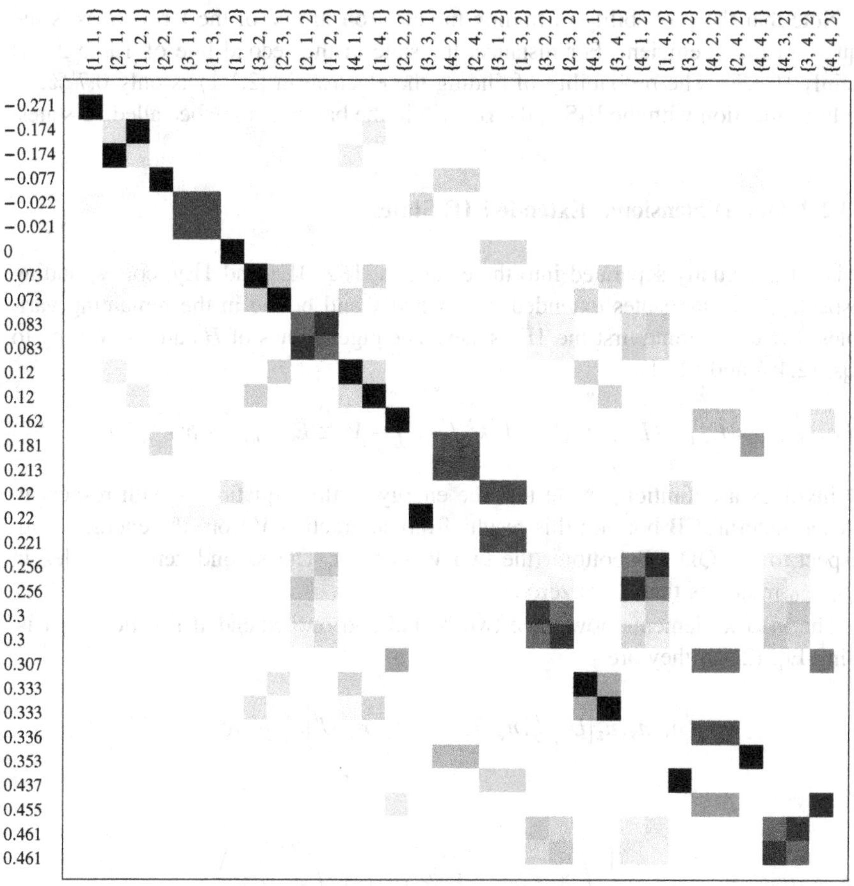

Fig. 2.7 Strength of the projection of the eigenstates corresponding to the exact Hamiltonian eigenenergy, to the *left*, on the H_0 Hamiltonian eigenstates (*above*). *Black* full projection; *white* zero projection. Eigenstates are characterized by their energy (*left*, in eV). Reproduced with permission. © 2013, Elsevier [18]

Table 2.5 Exact eigenfunctions for the bound states within the bandgap

Eigenenergy (eV)	Eigenstate
−0.271	$1.000\|1,1,1\rangle$
−0.174	$-0.083\|2,1,1\rangle + 0.996\|1,2,1\rangle - 0.015\|3,2,1\rangle - 0.026\|1,4,1\rangle$
−0.174	$0.996\|2,1,1\rangle + 0.083\|1,2,1\rangle - 0.015\|2,3,1\rangle - 0.026\|4,1,1\rangle$
−0.077	$-0.998\|2,2,1\rangle + 0.038\|4,2,1\rangle + 0.038\|2,4,1\rangle$
−0.022	$0.016\|1,1,1\rangle + 0.707\|3,1,1\rangle + 0.707\|1,3,1\rangle - 0.031\|3,3,1\rangle$
−0.021	$-0.707\|3,1,1\rangle + 0.707\|1,3,1\rangle$
0.000	$-0.997\|1,1,2\rangle + 0.056\|3,1,2\rangle + 0.056\|1,3,2\rangle$

Note that the probability of finding the electron in one of the term states is the square of the coefficient. For instance, the state in the second line of Table 2.5 is mainly $|1{,}2{,}1\rangle$. The probability of finding the electron in $|2{,}1{,}1\rangle$ is only 0.7 %.

In connection with the IBSC, the BS within the bandgap may be called IB states.

2.3.2.2 One-Dimensional Extended 1E States

This set is actually separated into three subsets, 1Ez, 1Ex and 1Ey, corresponding respectively to the states extended in z, x and y and bound in the remaining variables. Let us examine first the 1Ez states. The eigenvalues of H_0 are, according to Eqs. (2.46) and (2.41),

$$E_{0,n_x,n_y,k_{ez}} = E_{x,n_x} + E_{y,n_y} + \left(V + \hbar^2 k_{ez}^2/2m^*\right) - V \equiv E_{0,n_x,n_y,0} + \hbar^2 k_{ez}^2/2m^* \quad (2.57)$$

($\equiv$ involves a definition). Note that the energy in this equation is with respect to barrier material CB bottom; this results from subtracting V from the energies with respect to the QD CB bottom (the two Vs cancel). The second zero subindex in $E_{0,n_x,n_y,0}$ indicates that k_{ez} is zero.

The matrix elements now have two bound and one extended function; that is, using Eq. (2.52) they are

$$\left\langle \hat{n}_x, \hat{n}_y, \tilde{n}_z \middle| U' \middle| \hat{n}'_x, \hat{n}'_y, \tilde{n}'_z \right\rangle = \left\langle \hat{n}_x, \hat{n}_y \middle| U' \middle| \hat{n}'_x, \hat{n}'_y \right\rangle \delta_{\tilde{n}_z, \tilde{n}'_z}$$

where

$$\begin{aligned}\left\langle \hat{n}_x, \hat{n}_y \middle| U' \middle| \hat{n}'_x, \hat{n}'_y \right\rangle \equiv -V \times & \left[\int_{-a}^{a} \xi_{n_x}\xi_{n'_x}dx \left(\int_{-\infty}^{-a} \psi_{n_y}\psi_{n'_y}dy + \int_{a}^{\infty} \psi_{n_y}\psi_{n'_y}dy \right) \right. \\ & \left. + \left(\int_{-\infty}^{-a} \xi_{n_x}\xi_{n'_x}dx + \int_{a}^{\infty} \xi_{n_x}\xi_{n'_x}dx \right) \int_{-a}^{a} \psi_{n_y}\psi_{n'_y}dy \right] \\ & - 2V \times \left[\left(\int_{-\infty}^{-a} \xi_{n_x}\xi_{n'_x}dx + \int_{a}^{\infty} \xi_{n_x}\xi_{n'_x}dx \right) \left(\int_{-\infty}^{-a} \psi_{n_y}\psi_{n'_y}dy + \int_{a}^{\infty} \psi_{n_y}\psi_{n'_y}dy \right) \right]\end{aligned} \quad (2.58)$$

This matrix is independent of k_{ez}, which appears in Eq. (2.57).

In this case, the exact Hamiltonian matrix is,

$$\left\langle \hat{n}_x, \hat{n}_y, \tilde{n}_z \middle| H \middle| \hat{n}'_x, \hat{n}'_y, \tilde{n}'_z \right\rangle = \left\{ \left\langle n_x, n_y \middle| H_{0,0_z} \middle| n'_x, n'_y \right\rangle + \left\langle n_x, n_y \middle| U' \middle| n'_x, n'_y \right\rangle + \hbar^2 k_{ez}^2/2m^* \right\} \delta_{\tilde{n}_z, \tilde{n}'_z} \quad (2.59)$$

where $H_{0,0_z}$ is the diagonal matrix of elements equal to $E_{0,n_x,n_y,0}$. In this equation, k_e is linked to $\tilde{n}_z$ by Eq. (2.43) where, for large values of $\tilde{n}_z$, θ may be neglected. Otherwise, Eq. (2.42) is to be used.

The eigenvalues of $H_{0.0_z} + U'$ are represented in Fig. 2.6b in the sector labeled 1Ez. In total, there are 4 × 4 = 16 states; many of them degenerate. According to Eq. (2.55), they are the threshold of sets of continuous states rising upwards. This is clearly represented for the lower energy state by a light gray rectangle, but it should be noted that a separate continuum emanates from each state in this plot sector.

The development of the 16 2-dimensional (2D) eigenfunctions on the 16 $|n_x, n_y\rangle$ states is presented in the Supplemental Material of Ref. [18]. The 3D eigenfunctions are in this case the product of the 2D eigenfunctions times the 1D extended (in/out phase harmonic) wave corresponding to k_{ez}.

The positions of the thresholds obtained from the exact solution are compared to those when the Hamiltonian is H_0 (separation-of-variables-solution) in the Supplemental Material of Ref. [18]. The thresholds for H_0 are actually located much further above those of the exact solution; the perturbation potential brings them down substantially. However, the exact-solution thresholds approach, but do not invade, the barrier material bandgap. The 1st order approximation gives closer values to the exact solution (see Supplemental Material of Ref. [18]), but in most cases not very close.

For the 1Ex and 1Ey sets, an analogous treatment is developed. In this case, there are 4 × 2 = 8 $|n_y, n_z\rangle$ states and the same for $|n_x, n_z\rangle$. They appear in Fig. 2.6b and are the thresholds of continuous spectra of states. The development of the 2-dimensional eigenstates in $|n_y, n_z\rangle$ and the $E_{0,0,n_y,n_z}$ and $E_{1,0,n_y,n_z}$ approximations or, for the $|n_x, n_z\rangle$ state, $E_{0,n_x,0,n_z}$ and $E_{1,n_x,0,n_z}$ approximations, can also be found in the Supplemental Material of Ref. [18]. The first-order approximations are obtained, similarly to in Eq. (2.55), by adding the diagonal element of U' to the zero-order approximation.

2.3.2.3 Two-Dimensional Extended 2E States

This subsection deals with three sets, 2Exy, 2Eyz and 2Ezx, corresponding respectively to bound states only in z, x and y and extended in the remaining variables. Let us examine first the 2Exy states. The eigenvalues of H_0 are

$$\begin{aligned} E_{0,k_{ex},k_{ey},n_z} &= E_{z,n_x} + \left(V + \hbar^2 k_{ex}^2/2m^*\right) + \left(V + \hbar^2 k_{ey}^2/2m^*\right) - V \\ &\equiv E_{0,0,0,n_z} + \hbar^2 k_{ex}^2/2m^* + \hbar^2 k_{ey}^2/2m^* \end{aligned} \tag{2.60}$$

The subtracted V sets the energy origin at the barrier material CB bottom. The second and third zeroes in the subindex signify that k_{ex} and k_{ey} are zero.

The matrix elements now have one bound function and two extended; that is, they are, according to Eqs. (2.48) and (2.52),

$$\left\langle \tilde{n}_x, \tilde{n}_y, \hat{n}_z \middle| U' \middle| \tilde{n}'_x, \tilde{n}'_y, \hat{n}'_z \right\rangle = \langle \hat{n}_z | U' | \hat{n}'_z \rangle \delta_{\tilde{n}_x, \tilde{n}'_x} \delta_{\tilde{n}_y, \tilde{n}'_y}$$

where

$$\langle \hat{n}_z | U' | \hat{n}'_z \rangle \equiv -V \left[\int_{-a}^{a} \zeta_{n_z} \zeta_{n'_z} dz \right] - 2V \left[\left(\int_{-\infty}^{-a} \zeta_{n_z} \zeta_{n'_z} dx + \int_{a}^{\infty} \zeta_{n_z} \zeta_{n'_z} dx \right) \right] \quad (2.61)$$

This matrix is independent of the k_es, which appears in Eq. (2.60).

With the exact model for 2Exy states, the Hamiltonian matrix development is

$$\left\langle \tilde{n}_x, \tilde{n}_y, \hat{n}_z \middle| H \middle| \tilde{n}'_x, \tilde{n}'_y, \hat{n}'_z \right\rangle = \left\{ \langle n_z | H_{0,0_x,0_y} | n'_z \rangle + \langle n_z | U' | n'_z \rangle \right.$$
$$\left. + \hbar^2 k_{ex}^2 / 2m^* + \hbar^2 k_{ey}^2 / 2m^* \right\} \delta(k_{ex}, k'_{ex}) \delta(k_{ey}, k'_{ey}) \quad (2.62)$$

where $H_{0,0_x,0_y}$ is the diagonal matrix of elements of $E_{0,0,0,n_z}$. The eigenvalues of $H_{0,0_x,0_y} + U'$ are represented in Fig. 2.6b in the sector labeled 1Exy. For $|n_z\rangle$, there are only 2 states, which are the threshold of a set of continuous states rising upwards (medium gray rectangles). The development of the two 1D eigenfunctions on the two $|n_z\rangle$ states is presented in the Supplemental Material of Ref. [18]. The 3D eigenfunctions are in this case the product of the 1D eigenfunctions and two 1D extended (in/out phase harmonic) waves corresponding to the k_{ex} and k_{ey} vectors.

The same arguments may be applied to the 2Eyz and 2Ezx, where the bound states are $|n_x\rangle$ and $|n_y\rangle$ respectively. In this case, there are four states and four levels, which are represented in the corresponding sector of the Fig. 2.6b plot. Again the lowering of the $H_{0,0_y,0_z}$ or $H_{0,0_z,0_x}$ energy levels is very substantial, more than in the 1E sets; however, the levels do not penetrate into the bandgap, even though for the 2Eyz and 2Ezx sets they approach it closely. Details can be found in the Supplemental Material of Ref. [18].

2.3.2.4 3E Set of Three-Dimensional Extended States

The eigenvalues of H_0 are,

$$\begin{aligned} E_{0,k_{ex},k_{ey},k_{ez}} &= \left(V + \hbar^2 k_{ex}^2 / 2m^* \right) + \left(V + \hbar^2 k_{ex}^2 / 2m^* \right) + \left(V + \hbar^2 k_{ez}^2 / 2m^* \right) - V \\ &= 2V + \hbar^2 k_{ex}^2 / 2m^* + \hbar^2 k_{ey}^2 / 2m^* + \hbar^2 k_{ez}^2 / 2m^* \end{aligned} \quad (2.63)$$

The subtracted V sets the energy origin at the barrier material CB bottom. In addition, taking into account Eq. (2.52), the perturbation matrix becomes very simple because

$$\left\langle \tilde{n}_x, \tilde{n}_y, \hat{n}_z \middle| U' \middle| \tilde{n}'_x, \tilde{n}'_y, \hat{n}'_z \right\rangle = -2V\delta_{\tilde{n}_x,\tilde{n}'_x}\delta_{\tilde{n}_y,\tilde{n}'_y}\delta_{\tilde{n}_z,\tilde{n}'_z} \tag{2.64}$$

In consequence,

$$\left\langle \tilde{n}_x, \tilde{n}_y, \hat{n}_z \middle| H \middle| \tilde{n}'_x, \tilde{n}'_y, \hat{n}'_z \right\rangle = \left(\hbar^2 k_{ex}^2/2m^* + \hbar^2 k_{ey}^2/2m^* + \hbar^2 k_{ez}^2/2m^*\right)\delta_{\tilde{n}_x,\tilde{n}'_x}\delta_{\tilde{n}_y,\tilde{n}'_y}\delta_{\tilde{n}_z,\tilde{n}'_z} \tag{2.65}$$

which is the same as if the QD were absent although the wavevectors are not plane waves but the product of three 1D extended (in/out phase harmonic) wave corresponding to the k_{ex}, k_{ey} and k_{ez} vectors. Again, in this case, there is a substantial reduction of the H_0 eigenvalues: $2U$ for all of them; however, they do not penetrate into the bandgap, but reach its upper edge. In Fig. 2.6 they are represented by the dark gray zone.

In the box-shaped QD, the continuous spectrum is heavily degenerated (as is visible in Fig. 2.6) and a richer wavefunction structure is possible. For instance, a virtual bound state (0E) may be combined with 1E, 2E and 3E states with the same energy, giving a rather complex eigenfunction that is (as any eigenfunction) a stationary state.

2.4 Absorption Coefficients in Box Shaped Quantum Dots

The radiation (photon) absorption and emission mechanisms in the interaction between the light and a material system, whose foundations were attributed to Dirac by Fermi [21] in 1932, have been extensively presented in textbooks [3, 22, 23]. We have revisited this background for applications associated to solar cells [4]. In this section we follow Ref. [17] closely.

In device physics, the carrier generation rate per unit volume and time is given by the expression [24]

$$g = \int \alpha dF_{ph} = \int \frac{dF_{ph}}{dE}\alpha dE \tag{2.66}$$

where dF_{ph} is an elementary flux of photons per unit area and time in a narrow range of energies and α is the absorption coefficient.

For a single QD, the probability of photon absorption per unit time is given by the Fermi Golden Rule, that in the electric dipole approximation [3] (vol. II, p. 899) is (SI units)

$$w_{|\Xi;N\boldsymbol{q}\rangle\to|\Xi';(N-1)\boldsymbol{q}\rangle} = \frac{N_{ph,\boldsymbol{q}}}{\Omega}\frac{2\pi^2e^2E}{n_{ref}^2h\varepsilon_0}|\langle\Xi|\boldsymbol{r}\cdot\boldsymbol{\epsilon}|\Xi'\rangle|^2\delta(E_{\Xi'}-E_{\Xi}-E) \tag{2.67}$$

where $\boldsymbol{\epsilon}$ is the light polarization vector, n_{ref} is the refraction index of the medium, $N_{ph,\boldsymbol{q}}$ is the number of photons in the mode $\boldsymbol{q}$ of energy E and Ω is the crystal volume. This expression represents the number of excitons[5] generated per unit time due to absorption of photons in mode $\boldsymbol{q}$ (which, for the moment, includes the polarization state). It is required that there is an electron in state Ξ, and that state Ξ' is not filled with another electron which would exclude the transition. Furthermore, the spin in the initial and the final state has to be the same. As stated in Eq. (2.28), the matrix element $\langle\Xi|\boldsymbol{r}\cdot\boldsymbol{\epsilon}|\Xi'\rangle$ calculated from the exact wavefunction is the same as that calculated from the envelope function $\langle\Phi|\boldsymbol{r}\cdot\boldsymbol{\varepsilon}|\Phi'\rangle$.

If, in Eq. (2.67), we set

$$\sum_{\boldsymbol{q}} N_{ph,\boldsymbol{q}}/\Omega = n_{ph} \tag{2.68}$$

we are taking into account exciton generation by all modes with the same energy. The density of photons n_{ph} can be related to the photon flux by $F_{ph} = (c/n_{ref})n_{ph}$. Furthermore, if the fractional coverage of the surface with QDs is F_s and there are N_l QD layers per unit length, the density of QDs per unit volume is $F_sN_l/4ab$ where $4ab = 4a^2$ is the QD cross-section. With this nomenclature, the exciton generation rate per unit volume and time is

$$g = \int \frac{dF_{ph}}{dE}\frac{2\pi^2e^2E}{n_{ref}ch\varepsilon_0}\frac{|\langle\Xi|\boldsymbol{r}\cdot\boldsymbol{\epsilon}|\Xi'\rangle|^2}{4ab}F_sN_l\delta(E_{line}-E)dE \tag{2.69}$$

where $E_{line} = E_{\Xi'} - E_{\Xi}$. By analogy,

$$\alpha_{\Xi\to\Xi'}^{\max} = 2\frac{2\pi^2e^2E}{n_{ref}ch\varepsilon_0}\frac{|\langle\Xi|\boldsymbol{r}\cdot\boldsymbol{\epsilon}|\Xi'\rangle|^2}{4ab}F_sN_l\delta(E_{line}-E) \equiv \alpha'_{\Xi\to\Xi'}E\delta(E_{line}-E) \tag{2.70}$$

where α' is defined for convenience. The first factor of 2 is to account for the spin degeneracy of each state. Furthermore, the superindex *max* in the preceding formulas assumes that there is an electron in the initial state Ξ and none in the final state Ξ'. In general

$$\alpha_{\Xi\to\Xi'} = \alpha_{\Xi\to\Xi'}^{\max}f_{\Xi}(1-f_{\Xi'}) \tag{2.71}$$

---

[5] Exciton is an electron-hole pair linked by Coulombian attraction. At room temperature excitons are dissociated in a semiconductor. Electron-hole pairs and excitons are often considered synonymous, and is the case in this chapter.

where f_{Ξ} and $f_{\Xi'}$ are the filling factor (bounded to $0 < f < 1$) of the initial and final electron states. The filling factors are often described as Fermi factors of the type

$$f = \{\exp[(E - E_F)/kT] + 1\}^{-1} \tag{2.72}$$

where E is the energy of the state and E_F is the quasi Fermi level (QFL) for the state in consideration. The QFLs are usually different for the initial and final states except if the sample is in thermal equilibrium, in which case all converge to a single Fermi level.

In most cases, only the fundamental state $|1,1,1\rangle$ is filled with electrons to a significant extent. These electrons often come from donors introduced by doping. All the other states, including those in the CB, are essentially empty. However, this aspect will be further studied when dealing with the detailed balance conditions of the solar cells (Chap. 4). If the doping very strong, or under some other conditions, the first excited state may also become partly filled.

2.4.1 Influence of the Photon Polarization

The matrix element $\langle \Xi|\boldsymbol{\varepsilon}\cdot\boldsymbol{r}|\Xi'\rangle = \cos\phi\sin\theta\langle\Xi|x|\Xi\rangle + \sin\phi\sin\theta\langle\Xi|y|\Xi\rangle + \cos\theta\langle\Xi|z|\Xi'\rangle$ depends on the photon polarization, which is defined here by its Euler angles (φ, θ) with respect to the (x, y, z) axes. In the separation of variables approximation, the matrix element $\langle\Xi|x|\Xi'\rangle = \langle\Phi|x|\Phi'\rangle$ (see Eq. (2.28)) is the product three functions: one depending on x, one depending on y and one depending on z. To have $\langle\Phi|x|\Phi'\rangle \neq 0$ the quantum numbers n_y and n_z must be the same for Φ and Φ' because, otherwise, the internal product of the corresponding one-dimensional wavefunctions is zero, therefore only n_x can vary. What's more, as said before, the quantum numbers must be of different parity: otherwise $\langle\xi|x|\xi'\rangle = 0$. The same can be said for the other coordinates: the only non-zero matrix elements are those for which one single quantum number changes between two states of different parity.

For the SOTA QDs (Table 2.1), whose one-dimensional energies appear in Table 2.2, the matrix elements for polarized light in the direction of the changing QN ($\boldsymbol{\varepsilon}\cdot\boldsymbol{r} = x$ or y or z) can be calculated analytically [17]. They are given in Table 2.6.

Table 2.6 Bound to bound state transitions in the SOTA material for polarized light in the direction of the changing QN

Transition	n_x, n_y				n_z
	$1 \to 2$	$2 \to 3$	$3 \to 4$	$1 \to 4$	$1 \to 2$
E_{line} (eV)	0.100503	0.158622	0.175218	0.434343	0.311947
$\lvert\langle\Phi\lvert\epsilon\cdot\boldsymbol{r}\rvert\Phi'\rangle\rvert^2/4ab$	0.04889	0.03344	0.02868	1.9451×10^{-4}	0.01281

$U = 0.473$ eV; $m^* = 0.0294\ m_0$; $a = b = 8$ nm; $c = 3$ nm; $n_{ref} = 3.5$

As said before, only transitions of different parity in the polarization coordinate are permitted. Among these, the strongest transitions are between consecutive states.

With a general polarization, the square of the module of the matrix element, entering in the absorption coefficient, can be written as:

$$\begin{aligned}|\langle\Xi|\boldsymbol{\epsilon}\cdot\boldsymbol{r}|\Xi'\rangle|^2 =& \cos^2\varphi\sin^2\theta|\langle\Phi|x|\Phi'\rangle|^2+\sin^2\varphi\sin\theta^2|\langle\Phi|y|\Phi'\rangle|^2\\ &+\cos\theta^2|\langle\Phi|z|\Phi'\rangle|^2+2\cos\varphi\sin\phi\sin^2\theta\mathrm{Re}[\langle\Phi|x|\Phi'\rangle\langle\Phi|y|\Phi'\rangle]\\ &+2\cos\varphi\sin\theta\cos\theta\mathrm{Re}[\langle\Phi|x|\Phi'\rangle\langle\Phi|z|\Phi'\rangle]\\ &+2\sin\varphi\sin\theta\cos\theta\mathrm{Re}[\langle\Phi|y|\Phi'\rangle\langle\Phi|z|\Phi'\rangle]\end{aligned} \tag{2.73}$$

Since only one quantum number may vary, the terms containing two different matrix elements are always zero. For instance, if the changing quantum number is n_x, then $\langle\Phi|y|\Phi'\rangle$ is zero and so is the product $\langle\Phi|x|\Phi'\rangle\langle\Phi|y|\Phi'\rangle$. In consequence

$$|\langle\Xi|\boldsymbol{\epsilon}\cdot\boldsymbol{r}|\Xi'\rangle|^2 = \cos^2\varphi\sin^2\theta|\langle\Phi|x|\Phi'\rangle|^2+\sin^2\varphi\sin\theta^2|\langle\Phi|y|\Phi'\rangle|^2+\cos\theta^2|\langle\Phi|z|\Phi'\rangle|^2 \tag{2.74}$$

and only one of these three terms can be non-zero.

The incident photons have a multitude of polarizations and the observed data results from averaging $|\langle\Xi|\boldsymbol{r}\cdot\epsilon|\Xi'\rangle|^2$. In the case of normal illumination, since the electric field of the photons is transversal, the polarization vector has $\theta=\pi/2$ and $|\langle\Xi|\boldsymbol{r}\cdot\epsilon|\Xi'\rangle|^2=\cos^2\varphi|\langle\Phi|x|\Phi'\rangle|^2+\sin^2\varphi|\langle\Phi|y|\Phi'\rangle|^2$. Averaging over all the possible φ form 0 to 2π, we obtain

$$\left\langle|\langle\Xi|\epsilon\cdot\boldsymbol{r}|\Xi'\rangle|^2\right\rangle_z = \left(|\langle\Phi|x|\Phi'\rangle|^2+|\langle\Phi|y|\Phi'\rangle|^2\right)/2 \tag{2.75}$$

For in-plane illumination, the angle φ is fixed (normal to the propagation of incident light in the plane *x, y*) and averaging Eq. (2.74) over θ in the interval $(0,\pi)$ we obtain $\cos^2\varphi|\langle\Phi|x|\Phi'\rangle|^2/2+\sin^2\varphi|\langle\Phi|y|\Phi'\rangle|^2/2+|\langle\Phi|z|\Phi'\rangle|^2/2$ with only one term being non zero. For instance, if the quantum number that changes is n_y, then the squared matrix element is $\sin^2\varphi|\langle\Phi|y|\Phi'\rangle|^2/2$ with lobes in the $\varphi=\pm\pi/2$ that correspond to light beams directed along the x axis. In the case of in-plane illumination it may be that the polarization is fixed, usually either $\theta=0$ or $\theta=\pi/2$, and that the beam direction is also fixed. In this case Eq. (2.74) must be used without any averaging. In the case of isotropic illumination, as is the case for internal thermal photons emitted spontaneously by the semiconductor, the average is

$$\left\langle|\langle\Xi|\epsilon\cdot\boldsymbol{r}|\Xi'\rangle|^2\right\rangle_{isotropic} = |\langle\varphi|x|\varphi'\rangle|^2/3+|\langle\varphi|y|\varphi'\rangle|^2/3+|\langle\varphi|z|\varphi'\rangle|^2/3 \tag{2.76}$$

And, again, only one term is not zero.

Table 2.7 Polarization factor of some bound-to-bound state transitions for several polarization modalities (see nomenclature in text)

Initial	Final	*z*	*x*	*x/y*	*x/z*
111	121	1/2	1/2	1	0
	211	1/2	0	0	0
	141	1/2	1/2	1	0
	411	1/2	0	0	0
	112	0	1/2	0	1
121	131	1/2	1/2	1	0
	221	1/2	0	0	0
	421	1/2	0	0	0
	122	0	1/2	0	1
211	311	1/2	0	0	0
	221	1/2	1/2	1	0
	241	1/2	1/2	1	0
	212	0	1/2	0	1

Table 2.7 shows the factor that multiplies the element of matrix for all possible transitions between bound states under the assumption that only the fundamental level (1,1,1) and the first excited level with states (1,2,1) and (2,1,1) are appreciably filled with electrons. We consider four examples of illumination: an unpolarized beam in the z direction (denoted z), an unpolarized beam in the x direction (x), a y-polarized beam in the x direction (x/y) and a z-polarized beam in the x direction (x/z). The element of matrix that must be multiplied by this factor is the one whose final state QN differs from that of the initial state. For instance for the transition $|1,2,1\rangle \rightarrow |4,2,1\rangle$ the differing QN are $1 \rightarrow 4$ and the value of the element of matrix is 1.9451×10^{-4}, as found in Table 2.6. This number is to be multiplied by the factor in Table 2.7 corresponding to the beam direction and the polarization of the beam.

2.4.2 *Calculation of the Absorption Coefficients Between Bound States in the Separation-of-Variables Approximation*

The calculations to follow are based on the separation-of-variables approximation. The use of exact eigenfunctions will be discussed afterwards, and will support the use of this approximation.

We shall consider the cases in which the fundamental state and the first excited state are partially filled with electrons. All the other states are practically empty. Transitions to bound states are Dirac-delta shaped, following Eqs. (2.70) and (2.71). The absorption coefficient for transitions from the fundamental or first excited state to all other bound states is

$$\alpha^{BS} = f_{1,1,1} \sum_{(n_x,n_y,n_z)\neq(1,1,1)} (1 - f_{n_x,n_y,n_z}) E_{line} \alpha_{1,1,1}^{\prime n_x,n_y,n_z} \delta(E_{line} - E)$$
$$+ f_{1,2,1} \sum_{(n_x,n_y,n_z)\neq(1,1,1),(1,2,1),(2,1,1)} (1 - f_{n_x,n_y,n_z}) E_{line} \alpha_{1,2,1}^{\prime n_x,n_y,n_z} \delta(E_{line} - E)$$
$$+ f_{2,1,1} \sum_{(n_x,n_y,n_z)\neq(1,1,1),(1,2,1),(2,1,1)} (1 - f_{nx,ny,nz}) E_{line} \alpha_{2,1,1}^{\prime n_x,n_y,n_z} \delta(E_{line} - E) \tag{2.77}$$

where E_{line} is in each instance the energy difference between the final and initial state. The number of totally bound states in the QD is finite and usually restricted to few states. Note that $f_{1,2,1} = f_{2,1,1}$. In the α's, the subindices and superindices are the QNs of the initial and final states respectively.

The Dirac delta is in practice a narrow line whose width is controlled by the QDs' homogeneity. Small variations of size lead to line broadening. Broadening may also come from variations in the band offset or the effective mass. In practice, we use a Gaussian distribution such that

$$\delta(E - E_{line}) \cong \frac{1}{\sigma\sqrt{\pi}} \exp\left(-\frac{(E - E_{line})^2}{\sigma^2}\right) \tag{2.78}$$

In our calculations, we have set the energy dispersion $\sigma = 0.025$ eV.

Optimal filling for IB solar cells requires the fundamental state to be partially filled and the first excited state to be empty. For Quantum Dot Infrared Photodetectors (QDIPs) the fundamental state should be totally filled and the first excited state should be empty. Focusing our study on intermediate band solar cells, we present in Fig. 2.8 the absorption coefficients for the case (a) $f_{1,1,1} = 0.5$, $f_{1,2,1} = f_{2,1,1} = 0$ and for the case (b) $f_{1,1,1} = 1$, $f_{1,2,1} = f_{2,1,1} = 0.5$ corresponding to overdoping. The second might be the sought doping for QDIPs. The illumination conditions are specified in the figure caption.

Compared to the first case, the second case leads to a much stronger absorption and exhibits a peak at 0.159 eV that is absent in the first case. What is more, there is a reduction in the intensity of the first peak relative to the last peak. The different beam directions and polarizations have a strong influence on the absorption spectrum. In particular, the in-plane unpolarized illumination reduces the absorption by half at the 0.101 eV line with respect to the normal unpolarized illumination, but produces an additional absorption at the 0.312 eV line. It is to be stressed that there is strong influence of the filling state on the absorption coefficients. We might expect the peak at almost 0.1 eV to partially fill level (1,2,1) due to transitions from the level (1,1,1), and therefore the exclusive filling of the first level may not be realistic. The same can be said with the transitions of level (1,2,1) to level (2,2,1). A detailed balance analysis, to be presented later, will give us knowledge on the filling states and therefore a better vision of the absorption coefficients.

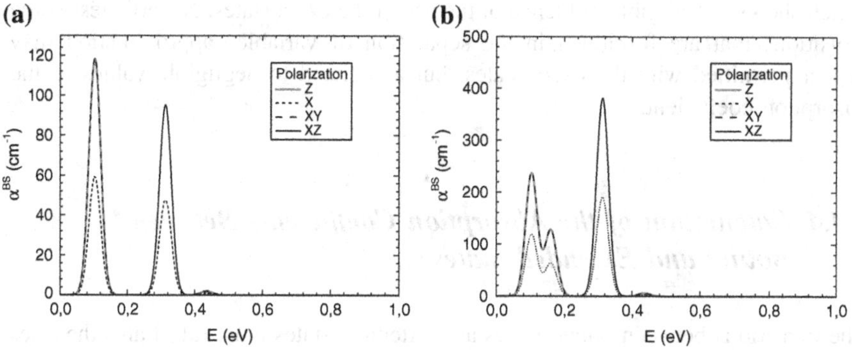

Fig. 2.8 Absorption coefficient due to transitions between bound states. *Grey solid line* z (normal) unpolarized beam; *black dotted line* x unpolarized beam; *black dashed line* y-polarized x-beam; *black solid line* z-polarized x-beam. **a** The fundamental (1,1,1) state is filled to 50 %. **b** The fundamental state is filled to 100 % and the first excited states (1,2,1) and (2,1,1) are filled to 50 %. $\sigma = 0.025$ eV. The 0.101 eV peak corresponds to (1,1,1) → (1,2,1), the 0.159 eV peak to (1,2,1) → (2,2,1), the 0.312 eV peak to (1,1,1) → (1,1,2) and the 0.434 eV peak to (1,1,1) → (1,4,2). QD data as in the SOTA prototype. Reproduced with permission. © 2013, Elsevier [25]

2.4.3 The Absorption Coefficients with Exact Eigenfunctions

One of the benefits of the separation of variables is the easy labeling of the states. The eigenfunctions in the exact solutions of Sect. 2.3.2 are clearly defined as linear combination of separable states but a concise labeling for them is difficult if it has to be illustrative.

Here, we label the states with their energies followed by (1) or (2) added to distinguish the first and second state listed in Table 2.5, when the states are degenerate. Using this table, that spans the exact states in the basis of separable states, and taking into account the forbidden transitions,

$$\begin{aligned} \langle -0.271|x|-0.174(1)\rangle &= -0.083\langle 1,1,1|x|2,1,1\rangle \\ \langle -0.271|y|-0.174(1)\rangle &= 0.996\langle 1,1,1|y|1,2,1\rangle \end{aligned} \tag{2.79}$$

For vertical unpolarized illumination the two preceding terms must be squared, added and divided by 2, as discussed in Sect. 2.4.1. According to Table 2.6, the values $\langle 1,1,1|x|2,1,1\rangle/4ab = \langle 1,1,1|y|1,2,1\rangle/4ab = 0.04889$ whereas $\langle 1,1,1|x|1,2,1\rangle/4ab = 0$ and $\langle 1,1,1|y|1,2,1\rangle/4ab = 0.04889$. Therefore

$$\begin{aligned} \left(\frac{|\langle -0.271|x|-0.174(1)\rangle|^2}{4ab} + \frac{|\langle -0.271|y|-0.174(1)\rangle|^2}{4ab}\right)\Big/ 2 = 0.02442 \\ \left(\frac{|\langle 1,1,1|x|1,2,1\rangle|^2}{4ab} + \frac{|\langle 1,1,1|y|1,2,1\rangle|^2}{4ab}\right)\Big/ 2 = 0.02445 \end{aligned} \tag{2.80}$$

which shows a negligible influence of the use of the exact states. Nevertheless some transitions that are forbidden in the separation of variables approximation may appear permitted with the exact states, but associated to negligible values of the absorption coefficients.

2.4.4 Calculation of the Absorption Coefficients Between Bound and Extended States

The transitions between bound states and extended states must fulfill also the rules of selection: initial and final states differ only in one QN and the (one-dimensional) parity of the initial and final states must be different. In consequence the final states have to belong to the type 1E, that is, they are extended only in one dimension.

We consider transitions from the fundamental or the first excited state, which are assumed to be the only states filled with electrons. We use the variable ρ to denote the extended dimension ($\rho = x$ or y or z). Only transitions to the extended states with parity different to the (one-dimensional) bound states are permitted. The separation between two such transitions of the same parity for large L, taking into account Eq. (2.43), which gives the values of the wavevector for a given QN, is,

$$\Delta k_{odd} = \Delta k_{even} = \pi/L. \tag{2.81}$$

Therefore, looking at the relationship between wavevectors and energies in Eq. (2.41), the density $N(E_{k_e;\rho})$ of extended states per unit of energy.

$$N(E_{k_e;\rho}) = \frac{1}{\Delta k_{even}}\frac{dk_e}{dE_{k_e;\rho}} = \frac{1}{\Delta k_{odd}}\frac{dk_e}{dE_{k_e;\rho}} = \frac{L(m^*)^{1/2}}{2^{1/2}\hbar\pi}E_{k_e;\rho}^{-1/2} \tag{2.82}$$

For the case $\rho = x$, the energy of the absorbed photons E_{line} is now

$$E_{line} = E_{0,k_e,n_y,n_z} - E_{0,n_x,n_y,n_z} = E_{k_e;x} - E_{n;x} \tag{2.83}$$

where $E_{k_e;x}$ and $E_{n;x}$ represent the one-dimensional energies in the x-direction. The same can be said for the other variables.

For the transition from a bound state to a state extended in one dimension we use the notation $\alpha'_{n;\rho}(k_e)$ where n is the initial value of the changing quantum number with corresponding coordinate ρ, and k_e corresponds to the final state wavevector outside the QD well, corresponding to the same coordinate.

The absorption coefficient for transitions from a bound state to all the 1E states is then

$$\begin{aligned}\alpha^{ES}_{n;\rho}(k_e) &= f_{n_x,n_y,n_z}\sum_k (1-f_C)E\alpha'_{n;\rho}(k_e)\delta(E_{line}-E)\\ &\cong f_{n_x,n_y,n_z}\int N(E_\rho)E\alpha'_{n;\rho}(k_e)\delta(E_{line}-E)dE\\ &= f_{n_x,n_y,n_z}N(E_{line}+E_{n;\rho})E_{line}\alpha'_{n;\rho}\left(k_e(E_{line}+E_{n;\rho})\right)\end{aligned} \tag{2.84}$$

Taking into account that only the fundamental and first excited states have appreciable probability of occupation, we can write

$$\alpha^{ES} = \sum_{\rho=x,y}\sum_{n=1,2} f_{n,1,1}\sum_{k_u} N(E_{line}+E_{n;\rho})E_{line}\alpha'_{n;\rho}\left(k_u(E_{line}+E_{n;\rho})\right) \tag{2.85}$$

where the sums must take into account all the combinations of possible transitions between bound states and the possible final extended state for each bound state.

There is a polarization influence in the calculation of the $\alpha'_{n;\rho}(k_e)$. It is presented in Table 2.8. The absorption coefficient is given by Eq. (2.85) summing over all the transitions found in this table. The matrix elements, which are analytic, are given in Ref. [17] (appendices). The $\alpha'_{u;n}$ are calculated with the absorption coefficient formula of Eq. (2.70).

The functions $\alpha^{ES}_{n,\rho}(k_e)$ defined in Eq. (2.85) are zero for photon energies below $E_{n;\rho}$. To account for this, we could multiply them by a step function starting at $E_{n;\rho}$. However, in the same way we have used a Gaussian function, instead of a Dirac delta, to account for the dispersion of the QDs size, here we can replace the step function with

$$H(E-E_{line}) \cong \left[\exp\left(-\frac{E-E_{u;n}}{2\sigma}\right)+1\right]^{-1} \tag{2.86}$$

Table 2.8 Polarization factor of some bound-to-extended state transitions for several polarization modalities (see nomenclature in text)

Initial	Final	z	x	x/y	x/z
111	1*k*1	1/2	1/2	1	0
	*k*11	1/2	0	0	0
	11*k*	0	1/2	0	1
121	1*k*1	1/2	1/2	1	0
	*k*21	1/2	0	0	0
	12*k*	0	1/2	0	1
211	2*k*1	1/2	1/2	1	0
	*k*11	1/2	0	0	0
	21*k*	0	1/2	0	1

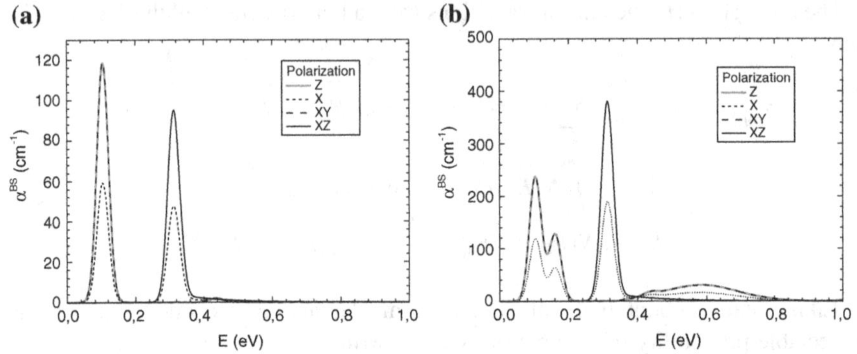

Fig. 2.9 Absorption coefficient due to transitions between bound to bound and bound to extended states. *Grey solid line* z (normal) unpolarized beam; *black dotted line* x unpolarized beam; *black dashed line* y-polarized x-beam; *black solid line* z-polarized x-beam. **a** The fundamental (1,1,1) state is filled to 50 %. **b** The fundamental state is filled to 100 % and the first excited states (1,2,1) and (2,1,1) are filled to 50 %. σ = 0.025 eV (This plot includes a correction of the results in [25] (which inspires this subsection) consisting on a wrong sign in a formula and an inaccurate normalization of the extended functions. We thank Aleksandr Panchak for calculations)

With $\sigma = 0.025$ eV. In Fig. 2.9 we present the absorption coefficients $\alpha^{BS} + \alpha^{ES}$ for the cases (a) $f_{1,1,1} = 0.5$, $f_{1,2,1} = f_{2,1,1} = 0$ and (b) $f_{1,1,1} = 1$, $f_{1,2,1} = f_{2,1,1} = 0.5$.

By comparing these plots with those in Fig. 2.8, we realize that the main part of the transitions is between bound states. Transitions to the extended states form tails of less importance. We insist here that the filling of states considered in this subsection may be unrealistic and a detailed balance analysis will be necessary for a realistic evaluation of the absorption. As a matter of fact, the absorptions studied in this subsection will fill the upper states leading to transference of the electrons from them to states of higher energy and ultimately to the CB. On the other hand, it is well known (and we shall see it later in this book) that absorption and emission of photons are strongly linked. This means that concurrent with the upwards traffic of electrons produced by the absorption of light there is a downwards traffic due to the emission of photons, stronger if the absorption coefficient is stronger and weaker if it weaker.

Additionally, we may observe that much of the absorption is produced at energies that are small. At room temperature the semiconductor contains an abundance of thermal photons at these energies. Often many more than those that can be injected by the sun inside the semiconductor. Table 2.9 shows the ratio of

Table 2.9 The ratio of externally incident to internal ambient photon density is also calculated for the line energies for AM0 illumination of 0.136 W · cm^{-2} (0.0121 W · cm^{-2} below 0.726 eV)

	n_x, n_y, $1 \to 2$	n_x, n_y, $2 \to 3$	n_x, n_y, $3 \to 4$	n_x, n_y, $1 \to 4$	n_z, $1 \to 2$
Energy (eV)	0.100503	0.158622	0.175218	0.434343	0.311947
External/internal ph. (1AM0 suns)	0.00009452	0.00054350	0.00091954	6.28179	0.088276

photons from an external AM0 source at the extraterrestrial radiation level (0.136 W cm^{-2}) and internal thermal photons at 300 K. For most levels, this ratio is, in the absence of concentration, well below one.

The consequence of this is that electrons may be pumped up, towards the CB, by these thermal photons. It might seem that this is very good; however, in reality, energy cannot be extracted from them unless the second law of thermodynamics is violated. The topic is discussed in [26]. In fact, we shall see that this strong up- and downwards communication between the IB levels and the CB via thermal photons prevents the splitting of their QFLs. This renders the IB levels a mere extension of the CB into the forbidden band and ultimately leads to a reduction in the open circuit voltage. The IB levels can be thus considered as ladder states for this communication between the IB states and the CB. The use of a host material of higher bandgap than the GaAs, such as the GaAlAs, may reduce the thermal escape and the drawbacks associated to it.

2.5 Experimental Measurement of the Intermediate Band to Conduction Band Absorption

The absorption studied in this chapter has been measured on several occasions [27–29]. In general, a waveguide like the one depicted in Fig. 2.10a is used to increase the optical path inside the QD material. An example of the measured absorption is presented in Fig. 2.10b.

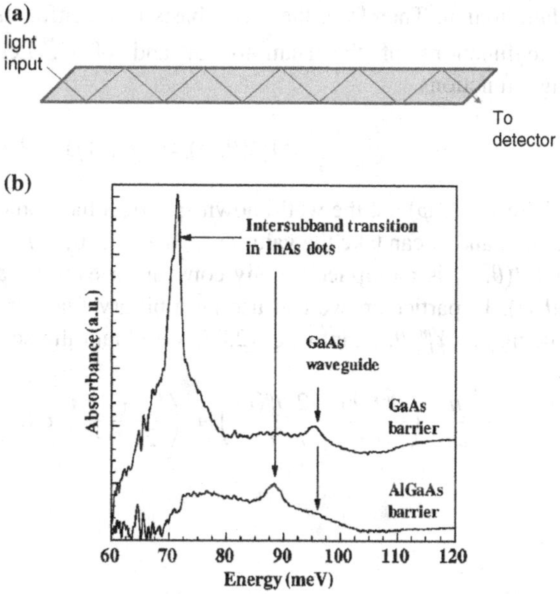

Fig. 2.10 Light absorption of low energy photons corresponding to the intraband transitions studied in this chapter. **a** 45° waveguide. **b** Absorbance for InAs QDs in GaAs and GaAlAs barrier materials. Reproduced with permission. © 1997 AIP [27]

Strong absorption is found at about 0.07 eV for the InAs/GaAs structure; this represents a photon energy below the one calculated in Fig. 2.9. This is not surprising since the position of the levels depends strongly on the QD size. Other experiments show the main peak at around 0.15 eV [28]. Also, the barrier material, which affects to the offset potential, affects the position of the absorption lines.

2.6 Spherical Quantum Dots

2.6.1 Central Potential Hamiltonian

In this section we are going to calculate the energy spectrum and the eigenfunctions of a QD with spherical shape and square potential.

In spherical coordinates, the envolvent TISE is [3]

$$\begin{aligned}
&\left(\frac{p_r^2}{2m^*}+\frac{\hat{l}^2}{2m^*r^2}+U(r)\right)\Phi(r,\theta,\varphi)=E\Phi(r,\theta,\varphi)\\
&p_r\equiv\frac{\hbar}{i}\frac{1}{r}\frac{\partial}{\partial r}r\\
&\hat{l}^2\equiv-\frac{\hbar^2}{\sin^2\theta}\left[\sin\theta\frac{\partial}{\partial\theta}\left(\sin\theta\frac{\partial}{\partial\theta}\right)+\frac{\partial^2}{\partial\varphi^2}\right]
\end{aligned}\tag{2.87}$$

$\hat{\boldsymbol{l}}$ has a physical meaning; it is the kinetic moment $\boldsymbol{r}\times\boldsymbol{p}$.

It is visible that $\hat{l}^2$ (only depending on angular variables) commutes with the Hamiltonian. Therefore, there is a basis of eigenfunctions that are simultaneously eigenfunctions of the Hamiltonian and of $\hat{l}^2$. Let us look first for the $\hat{l}^2$ eigenfunctions.

$$\hat{l}^2Y_l^m(\theta,\varphi)=l(l+1)Y_l^m(\theta,\varphi)\tag{2.88}$$

where $Y_l^m(\theta,\varphi)$ are the well-known spherical harmonics; l can take the values 0, 1, 2, 3,… and m can take the values $-l, -l+1,\ldots 0,\ldots l-1, l$. The equation is verified if $Y_l^m(\theta,\varphi)$ is multiplied by any constant or even by a function not depending on (θ,φ). In particular, we can use an arbitrary function of r such as $f(r)$. By introducing $f(r)Y_l^m(\theta,\varphi)$ into Eq. (2.87) we obtain the so called radial equation,

$$-\frac{\hbar^2}{2m^*}\left(\frac{\partial^2 f(r)}{\partial r^2}+\frac{2}{r}\frac{\partial f(r)}{\partial r}\right)+\left(\frac{l(l+1)}{2m^*r^2}+U(r)-E\right)f(r)=0\tag{2.89}$$

Table 2.10 Some expanded spherical Bessel functions

l	$j_l(kr)$	$-n_l(kr)$	$-i^l h_l^{(+)}(i\kappa r)$
0	$\frac{\sin(kr)}{kr}$	$-\frac{\cos(kr)}{kr}$	$-\frac{\exp(-\kappa r)}{kr}$
1	$-\frac{\cos(kr)}{kr}+\frac{\sin(kr)}{(kr)^2}$	$-\frac{\cos(kr)}{kr}-\frac{\sin(kr)}{(kr)^2}$	$\frac{i\exp(-\kappa r)(1+\kappa r)}{(\kappa r)^2}$
2	$-\frac{3\cos(kr)}{(kr)^2}+\frac{\left(3-(kr)^2\right)\sin(kr)}{(kr)^3}$	$-\frac{3\sin(kr)}{(kr)^2}-\frac{\left((3-(kr)^2\right)\cos(kr)}{(kr)^3}$	$\frac{\exp(-\kappa r)\left(3+3\kappa r+(\kappa r)^2\right)}{(\kappa r)^4}$
3	$\frac{\left(-15+(kr)^2\right)\cos(kr)}{(kr)^3}-\frac{\left(-15+6(kr)^2\right)\sin(kr)}{(kr)^4}$	$\frac{\left(-15+(kr)^2\right)\sin(kr)}{(kr)^3}+\frac{\left(-15+6(kr)^2\right)\cos(kr)}{(kr)^4}$	$-i\exp(-\kappa r)\times\frac{\left(15+15\kappa r+6(\kappa r)^2+(\kappa r)^3\right)}{(\kappa r)^4}$

In the case of a constant potential $U(r) = V_0$

$$\left(\frac{\partial^2 f(kr)}{\partial(kr)^2}+\frac{2}{r}\frac{\partial f(kr)}{\partial(kr)}\right)+\left(1-\frac{l(l+1)}{(kr)^2}\right)f(kr)=0 \qquad (2.90)$$
$$\text{with} \quad k=\frac{\sqrt{2m^*(E-V_0)}}{\hbar}$$

This is the spherical Bessel equation. Its general solution is a linear combination of two particular solutions, e.g. the functions $j_l(kr)$ and $n_l(kr)$ (see definitions in [3, Appendix B]) whose asymptotic behavior for kr $\rightarrow\infty$ is $\sin(kr - l\pi/2)/kr$ and $\cos(kr - l\pi/2)/kr$ respectively. However, only $j_l(kr)$ is acceptable; at the origin $j_l(kr)\rightarrow(kr)^l$, whereas $n_l(kr)$ diverges at the origin as $n_l(kr)\rightarrow(kr)^{-l-1}$, which is not acceptable. The functions $h^{(+)}\equiv n_l + ij_l$ and $h^{(-)}\equiv n_l - ij_l$, with asymptotic behavior $\exp(i(kr - l\pi/2))/kr$ and $\exp(-i(kr - l\pi/2))/kr$ are also acceptable in some cases.

For a given l index, the spherical Bessel functions may be expanded into more common functions. We present in Table 2.10 these expansions for some values of l. A drawing of some of these functions is presented in Fig. 2.11.

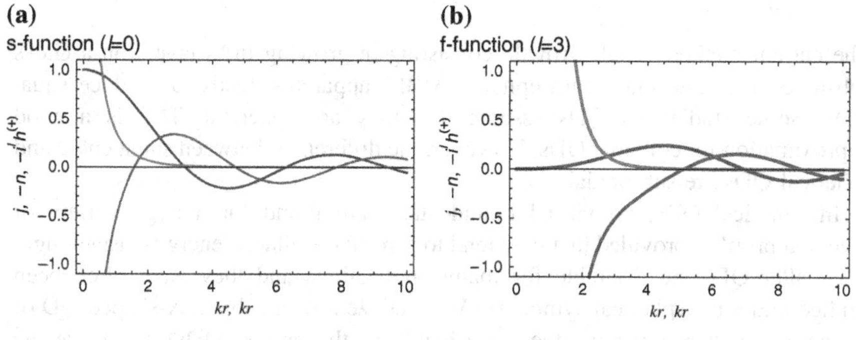

Fig. 2.11 Some spherical Bessel functions. *Blue*, $j_l(kr)$; *red*, $-n_l(kr)$; *yellow*, $-i^l h_l^{(+)}(i\kappa r)$ (color figure online)

2.6.2 Quantum Dots of Spherical Symmetry and Square Potential

If now we consider the case of a square potential, such that $V_0 = -V$ for $r \leq a$ and $V_0 = 0$ for $r > a$ (the origin of energies has been put at the host semiconductor CB) we have two cases:

For $E < 0$, we can define $i\kappa \equiv k$ so that $\kappa = (-2\ m^* E)^{1/2}$. Outside the QD ($r > a$) only the solutions with an exponentially decreasing behavior can be accepted; if this is the case, the only acceptable solution is $h^{(+)}(kr) = h^{(+)}(i\kappa r)$ (with an arbitrary constant); inside the QD ($r \leq a$), $k_i \equiv (2\ m^*(E + V))^{1/2}$ and only the solution $j_l(k_i r)$ is acceptable (with an arbitrary constant). Both have to coincide ar $r = a$ as well as their fist derivative. Therefore the condition

$$\frac{dh^+(i\kappa r)}{dr} \Big/ h^+(i\kappa r) = \frac{dj(k_i r)}{dr} \Big/ j(k_i r) \tag{2.91}$$

has to be fulfilled. This leads to a set of discrete values for the energy. Note that the Bastard boundary conditions of Sect. 2.2.2 can also be applied to this case.

For $E > 0$, is $C_1 j_l\,(k_i r)$ inside the QD and $C_2[\cos(\delta_l) j_l\,(k_e r) + \sin(\delta_l) C_2 n_l(k_e r)]$ with C_1, C_2 and δ_l three arbitrary constants and $k_e \equiv (2\ m^* E)^{1/2}$. The matching of function and derivative lead to

$$k_e \frac{d(\cos(\delta_l) j_l(k_e r) + \sin(\delta_l) n_l(k_e r))}{dr} \Big/ (\cos(\delta_l) j_l(k_e r) + \sin(\delta_l) n_l(k_e r))$$
$$= k_i \frac{dj(k_i r)}{dr} \Big/ j(k_i r) \tag{2.92}$$

and this determines δ_l. Note that, in this case, any value of E is valid and the spectrum is continuous.

2.6.3 Spherical Symmetry and the Cubic Case

The current method of QD growth, consisting in growing InAs layers on a GaAs substrate in a molecular beam epitaxy (MBE) apparatus, tends to produce squat QDs. Some studies on QDs assume that they are spherical. This is a good approximation for colloidal QDs. However, the differences between squat cubic and spherical QDs are substantial.

In spherical QDs, no virtual bound states are found for a large variety of potential profiles, provided that they tend to zero (the wellhead energy) fast enough.

Smaller QDs are desirable for many applications and they have often been studied under the spherical symmetry. We analyze now a cubic box-shaped QD of dimensions $6 \times 6 \times 6\ \text{nm}^3$ (the same height of the present QDs) and leave the

Table 2.11 Eigenvalues E_0 of the separation-of-variables Hamiltonian H_0 presented in increasing order and 1st order approximations E due to the perturbation potential V' for wavefunctions of the class 0E (bound states). The energy origin is located at the barrier material CB bottom. The 1st order approximation is exact

State	$\|1,1,1\rangle$	$\|2,1,1\rangle$	$\|1,2,1\rangle$	$\|1,1,2\rangle$	$\|2,2,1\rangle$	$\|2,1,2\rangle$	$\|1,2,2\rangle$	$\|2,2,2\rangle$
E_0 (eV)	−0.0681	0.2105	0.2105	0.2105	0.4891	0.4891	0.4891	0.7677
E (eV)	−0.0890	0.1385	0.1385	0.1385	0.2767	0.2767	0.2767	0.3728

remaining parameters unchanged. For the bound states we obtain Table 2.11. In this case, the 1st order approximation, studied in Sect. 2.3, coincides with the exact solution. The exact eigenvectors are the separation-of-variables solutions denoted in the table.

A spherical QD with the same volume as the $6 \times 6 \times 6$ nm^3 cube has a diameter of 7.4 nm and the s-state ($l = 0$) energy is −0.0902 eV, rather close to the only below-bandgap bound state energy of −0.0890 eV obtained for the cube. In the sphere this is the only bound state; neither a second s-state nor states with any other angular symmetry are present for the parameters used. Thus, the absence of virtual bound states is a singularity of the high symmetry of the spherical QD.

In the cube, there are also eigenfunctions that are extended in one dimension or in two dimensions and bound in the remaining ones. Their continuous spectrum starts well inside the host material CB. Values are not reported for lack of specific interest. These types of solutions are also missing in the spherical QDs.

Apparently, the spherical-wave solutions are missing in the cube. Actually this is not true. A spherical wave can be built as continuous linear combination of plane waves. All the $\boldsymbol{k}$-vectors of module k are the components. In effect, let us analyze the function

$$\Phi(\boldsymbol{r}) = \int \delta(|\boldsymbol{k}| - k) \exp(i\boldsymbol{k} \cdot \boldsymbol{r}) d^3 k = 2\pi k^2 \int_0^\pi \exp(ikr \cos \vartheta) \sin \vartheta d\vartheta = \frac{4\pi k}{r} \sin(kr) \tag{2.93}$$

where ϑ is the angle between the vectors $\boldsymbol{k}$ and $\boldsymbol{r}$ (spherical coordinates in the $\boldsymbol{k}$-space are used and the polar-axis is taken to be in the direction of $\boldsymbol{r}$). For the final result, the change of variables $\cos \vartheta = u$ has been made.

For large values of r, the function $\exp(i\boldsymbol{k} \cdot \boldsymbol{r})$ is also an eigenstate of the box-shaped potential well, as is the function $\exp(-i\boldsymbol{k} \cdot \boldsymbol{r})$. To prove this, let us go back to the one dimensional eigenfunctions,

$$\begin{aligned} e^{ik_x x} &= \frac{(e^{i\theta_o}\xi_{k,e} + ie^{i\theta_e}\xi_{k,o})}{\cos(\theta_o - \theta_e)} \\ e^{-ik_x x} &= \frac{(e^{-i\theta_o}\xi_{k,e} - ie^{-i\theta_e}\xi_{k,o})}{i\sin(\theta_e - \theta_o)} \end{aligned} \tag{2.94}$$

In the first octant (where all the space coordinates are positive) the exponential is developed in the eigenfunctions described above as (e and o mean even and odd respectively)

$$
\begin{aligned}
\mathrm{e}^{i\boldsymbol{k}\cdot\boldsymbol{r}} = & \frac{\mathrm{e}^{i(\theta_{xo}+\theta_{yo}+\theta_{zo})}\left|k_x e, k_y e, k_z e\right\rangle + i\mathrm{e}^{i(\theta_{xo}+\theta_{yo}+\theta_{ze})}\left|k_x e, k_y e, k_z o\right\rangle}{\cos(\theta_{xo}-\theta_{xe})\cos(\theta_{yo}-\theta_{ye})\cos(\theta_{zo}-\theta_{ze})} \\
& + \frac{i\mathrm{e}^{i(\theta_{xo}+\theta_{ye}+\theta_{zo})}\left|k_x e, k_y o, k_z e\right\rangle + i\mathrm{e}^{i(\theta_{xe}+\theta_{yo}+\theta_{zo})}\left|k_x o, k_y e, k_z e\right\rangle}{\cos(\theta_{xo}-\theta_{xe})\cos(\theta_{yo}-\theta_{ye})\cos(\theta_{zo}-\theta_{ze})} \\
& - \frac{\mathrm{e}^{i(\theta_{xo}+\theta_{ye}+\theta_{ze})}\left|k_x e, k_y o, k_z o\right\rangle + \mathrm{e}^{i(\theta_{xe}+\theta_{yo}+\theta_{ze})}\left|k_x o, k_y e, k_z o\right\rangle}{\cos(\theta_{xo}-\theta_{xe})\cos(\theta_{yo}-\theta_{ye})\cos(\theta_{zo}-\theta_{ze})} \\
& - \frac{\mathrm{e}^{i(\theta_{xe}+\theta_{ye}+\theta_{zo})}\left|k_x o, k_y o, k_z e\right\rangle i\mathrm{e}^{i(\theta_{xe}+\theta_{ye}+\theta_{ze})}\left|k_x o, k_y o, k_z o\right\rangle}{\cos(\theta_{xo}-\theta_{xe})\cos(\theta_{yo}-\theta_{ye})\cos(\theta_{zo}-\theta_{ze})}
\end{aligned}
\tag{2.95}
$$

In other octants, the linear combination is somewhat different insofar as the de-phasing angles change the sign for the negative coordinate. For the wave plane travelling backwards, a slightly different linear combination is to be built.

Once the plane wave is built, a linear combination of them leads to the spherical wave of Eq. (2.93). Notice that, in the context of this paper, the $\boldsymbol{k}$s are restricted to the first octant in the $\boldsymbol{k}$-space. However, the same result is obtained if the linear combination $\exp(i\boldsymbol{k}\cdot\boldsymbol{r}) + \exp(-i\boldsymbol{k}\cdot\boldsymbol{r})$ is integrated.

As is well known, plane waves can also be obtained as a linear combination of spherical waves (see e.g., [3]). Thus, the main difference between the spherical and box-shaped square-potential-well QDs is that the latter presents virtual 0E bound states and linear and planar 1E and 2E extended states, all absent in the spherical QD. It has to be stressed that the solutions we are presenting here are exact solutions for the model described. The same model is used for the spherical QD with only the change of shape.

2.7 Concluding Remarks of This Chapter

In this chapter, we have described the effective mass Schrödinger equation to calculate the envelope functions used to describe the eigenfunctions and calculate the eigenvalues of nanostructured semiconductors in the cases where a single band envelope equation is a good approximation. This is usually the case of the conduction band eigenfunctions in zincblende semiconductors. The requirement is that the function $u_{v,\boldsymbol{k}}$ is independent of $\boldsymbol{k}$ at least for small values of this parameter.

In our calculations, a box-shaped QD is studied in detail. The CB band offset between QD and host is assumed square (for comparison with more accurate calculations, look at [30, 31]). The QD is assumed to be box shaped. We believe that these two assumptions are a reasonable representation of the real QDs in zincblende semiconductors, and indeed easier to treat. A separation-of-variables solution of the box shape allows an easy labeling of the eigenstates thorough a set of three QNs corresponding to the one-dimensional eigenfunctions, the eigenvalues being the sum of the one-dimensional eigenvalues. An exact solution of the mathematical problem of the box-shaped square potential, obtained by diagonalization of this Hamiltonian in the basis of separation-of-variables eigenfunctions, confirm the existence of virtual bound states that are sometimes described as resonances of extended wavefunctions and whose existence is even a matter of doubt by some authors [32, 33] based on mathematical grounds. In this chapter, they are very clearly proven, also based on mathematical grounds, and play a very important role in the light absorption processes.

It is also found that the separation-of-variables energy spectrum (the eigenvalues) is accurate for the IB states and for the lower energy virtual bound states, but it is not acceptable for the higher energy virtual bound states. It is exact for extended states. A first-order energy approximation is reasonably accurate for some applications in the case of high energy virtual bound states.

In the nanostructured semiconductors, we use at least two different semiconductors, the barrier or host semiconductor and the nanostructure semiconductor. As implied in the preceding text, much of the potential in the nanostructured material is due to the CB offset between host and nanostructure. But also the effective masses are different in the nanostructure and the host. As a rule, this is not taken into consideration in our studies. We generally use the QD effective mass everywhere. We also recommended using the host material effective mass for the study of extended states (although this recommendation is not followed in the examples presented). There is also the possibility of using a Hermitical effective mass energy operator that applies for variable effective mass. The reason we do not use it is because it is not theoretically compatible with our basic assumption of a position-variable energy-dispersion function, though it must be mentioned that a position-variable energy-dispersion function is per se a doubtful concept. However, leaving aside the doubtfulness of the concept, in the separation of variables approximation, using the boundary conditions at the QD edges implied by the variable-effective-mass energy operator (the Bastard boundary conditions) leads to a bound-state energy spectrum that is close that obtained using the QD effective mass for the low energies and close to that obtained using the host material effective mass for the high energies. In summary, it seems that the eigenvalues and probably the eigenfunctions obtained using a position-variable effective mass are more accurate than using the approach we propose. However, to verify this statement, it is necessary to find the solution of the exact equation in the framework presented in this book, which is more difficult and has not been undertaken so far. The same can be said about the light absorption properties resulting for the use of this approximation.

A very interesting conclusion of this chapter is that the absorption properties, in the one band effective mass approximation, depend only on the envelope functions and these depend on the material only through the offset potential and the effective mass. Otherwise they depended mainly on the shape and size of the QD. This is very good because the knowledge of the periodic part of the Bloch function is a serious problem. This statement is only true if the integral multiplication rule can be applied and would not be true for too small QDs.

In this context, the workhorse of the absorption of light is produced between bound states in the IB and virtual bound states in the CB. One-dimensional extended states, a modality specific to the box-shaped QDs, play a secondary role in the light absorption. Spherical QDs have neither virtual bound states nor states which are partially bound and partially extended (like the one dimensional extended states). In conclusion, it is probable that they absorb the light more weakly than the extended states, but this cannot be ascertained without a specific calculation.

Experimental confirmation of the calculations of this chapter is also presented, although in structures we have not analyzed theoretically so that any attempt of quantitative confirmation is not possible.

Finally, it seems that the large QDs produced in the Stransky Krastanov growth mode leads to excited states of the IB that, at least in the InAs/GaAs QD/host system, lead to what we call ladder states able to absorb the abundant thermal photons and putting in thermal contact IB and CB. This results in a reduction of the solar cells open circuit voltage. In fact, this reduction totally disappears at low temperatures [9, 34] at which the bandgap voltage is almost fully achieved.

References

1. Datta S (1989) Quantum phenomena. Molecular series on solid state devices, vol 8. Addison Wesley, Reading
2. Kaxiras E (2003) Atomic and electronic structure of solids. Cambridge University Press, New York
3. Messiah A (1960) Mécanique quantique. Dunod, Paris
4. Luque A, Marti A, Mendes MJ, Tobias I (2008) Light absorption in the near field around surface plasmon polaritons. J Appl Phys 104(11):113118. doi:10.1063/1.3014035
5. Luque A, Antolín E, Linares PG, Ramiro I, Mellor A, Tobías I, Martí A (2013) Interband optical absorption in quantum well solar cells. Sol Energy Mater Sol Cells 112:20–26. doi:10.1016/j.solmat.2012.12.045
6. Bastard G (1990) Wave mechanics applied to semiconductor nanostructures. Monographies de Physique. Les Editions de Physique, Paris
7. Antolín E, Marti A, Farmer CD, Linares PG, Hernández E, Sánchez AM, Ben T, Molina SI, Stanley CR, Luque A (2010) Reducing carrier escape in the InAs/GaAs quantum dot intermediate band solar cell. J Appl Phys 108(6):064513
8. Martí A, Antolín E, Cánovas E, López N, Luque A, Stanley C, Farmer C, Díaz P, Christofides C, Burhan M (2006) Progress in quantum-dot intermediate band solar cell research. In: Poortmans J, Ossenbrink H, Dunlop E, Helm P (eds) Proceedings of the 21st European photovoltaic solar energy conference. WIP-Renewable Energies, Munich, pp 99–102
9. Linares PG, Marti A, Antolin E, Farmer CD, Ramiro I, Stanley CR, Luque A (2012) Voltage recovery in intermediate band solar cells. Sol Energy Mater Sol Cells 98:240–244. doi:10.1016/j.solmat.2011.11.015
10. Luque A, Marti A, Antolín E, Linares PG, Tobias I, Ramiro I (2011) Radiative thermal escape in intermediate band solar cells. AIP Adv 1:022125
11. Antolín E, Martí A, Luque A (2011) The lead salt quantum dot intermediate band solar cell. Paper presented at the 37th photovoltaic specialists conference, Seattle, pp 001907–12

12. Linares PG, Marti A, Antolin E, Luque A (2011) III-V compound semiconductor screening for implementing quantum dot intermediate band solar cells. J Appl Phys 109:014313
13. Popescu V, Bester G, Hanna MC, Norman AG, Zunger A (2008) Theoretical and experimental examination of the intermediate-band concept for strain-balanced (In,Ga)As/Ga(As,P) quantum dot solar cells. Phys Rev B 78:205321
14. Lazarenkova OL, Balandin AA (2001) Miniband formation in a quantum dot crystal. J Appl Phys 89(10):5509–5515
15. Shao Q, Balandin AA, Fedoseyev AI, Turowski M (2007) Intermediate-band solar cells based on quantum dot supracrystals. Appl Phys Lett 91(16):163503. doi:10.1063/1.2799172
16. Luque A, Marti A, Antolin E, Garcia-Linares P (2010) Intraband absorption for normal illumination in quantum dot intermediate band solar cells. Sol Energy Mater Sol Cells 94:2032–2035
17. Luque A, Marti A, Mellor A, Marron DF, Tobias I, Antolín E (2013) Absorption coefficient for the intraband transitions in quantum dot materials. Prog Photovoltaics 21:658–667. doi:10.1002/pip.1250
18. Luque A, Mellor A, Tobías I, Antolín E, Linares PG, Ramiro I, Martí A (2013) Virtual-bound, filamentary and layered states in a box-shaped quantum dot of square potential form the exact numerical solution of the effective mass Schrödinger equation. Phys B 413:73–81. doi:10.1016/j.physb.2012.12.047
19. Bastard G, Ziemelis UO, Delalande C, Voos M, Gossard AC, Wiegmann W (1984) Bound and virtual bound-states in semiconductor quantum wells. Solid State Commun 49(7):671–674
20. Ferreira R, Bastard G (2006) Unbound states in quantum heterostructures. Nanoscale Res Lett 1(2):120–136. doi:10.1007/s11671-006-9000-1
21. Fermi E (1932) Quantum theory of radiation. Rev Mod Phys 4(1):0087–0132
22. Heitler W (1944) The quantum theory of radiation. Oxford University Press, Oxford
23. Berestetski V, Lifchitz E, Pitayevski L (1972) Théorie quantique relativiste. Mir, Moscou
24. Luque A, Marti A, Antolín E, Linares PG, Tobías I, Ramiro I, Hernandez E (2011) New Hamiltonian for a better understanding of the quantum dot intermediate band solar cells. Sol Energy Mater Sol Cells 95:2095–2101. doi:10.1016/j.solmat.2011.02.028
25. Luque A, Marti A, Mellor A, Marron DF, Tobias I, Antolín E (2013) Absorption coefficient for the intraband transitions in quantum dot materials. Prog Photovoltaics 21:658–667. doi:10.1002/pip.1250
26. Luque A, Martí A, Cuadra L (2001) Thermodynamic consistency of sub-bandgap absorbing solar cell proposals. IEEE Trans Electron Devices 48(9):2118–2124
27. Phillips J, Kamath K, Zhou X, Chervela N, Bhattacharya P (1997) Photoluminescence and far-infrared absorption in Si-doped self-organized InAs quantum dots. Appl Phys Lett 71(15):2079–2081. doi:10.1063/1.119347
28. Sauvage S, Boucaud P, Julien FH, Gerard JM, ThierryMieg V (1997) Intraband absorption in n-doped InAs/GaAs quantum dots. Appl Phys Lett 71(19):2785–2787. doi:10.1063/1.120133
29. Durr CS, Warburton RJ, Karrai K, Kotthaus JP, Medeiros-Ribeiro G, Petroff PM (1998) Interband absorption on self-assembled InAs quantum dots. Phys E 2(1–4):23–27. doi:10.1016/s1386-9477(98)00005-8
30. Popescu V, Bester G, Zunger A (2009) Strain-induced localized states within the matrix continuum of self-assembled quantum dots. Appl Phys Lett 95(2):023108. doi:10.1063/1.3159875
31. Pryor C (1998) Eight-band calculations of strained InAs/GaAs quantum dots compared with one-, four-, and six-band approximations. Phys Rev B 57(12):7190–7195
32. Kato T (1959) Growth properties of solutions of the reduced wave equation with a variable coefficient. Commun Pure Appl Math 12:403–425
33. Agmon S (1970) Lower bounds for solutions of the Schrödinger equations. J Anal Math 23:1–25
34. Luque A, Marti A, Stanley C (2012) Understanding intermediate-band solar cells. Nat Photonics 6(3):146–152. doi:10.1038/nphoton.2012.1

Chapter 3
A Four Band Approximation: The Empiric k·p Hamiltonian

Abstract The grounds of the multiband envelope equations are described and proven here. Eigenfunctions of the nanostructured semiconductor are herein described as linear combinations of these envelope functions multiplied by the corresponding periodic parts of Bloch functions. The later are the Bloch functions of a homogeneous semiconductor of reference calculated at the Γ point in several bands. This treatment is called the k·p model. The preceding general theory is applied to the case of zincblende materials where the relevant bands are the conduction band, two valence bands of light and heavy holes and a third valence band split off from the others by the spin-orbit interaction. In a first step, the spin-orbit interaction is neglected leading to a simple four band k·p model (the zero order approximation) in which each periodic part of a Bloch function is spin degenerated and the equations for spin-up and for spin-down are the same. The spin-orbit interaction as well as the strain perturbation created by the inclusion of the quantum dots is introduced by using the experimental effective mass dispersion equations with different effective masses for each one of the four bands used. This is the so called Experimental k·p Hamiltonian (EKPH) model. By diagonalization of this Hamiltonian, the full spectrum of eigenenergies of a quantum-dot nanostructured semiconductor is obtained, and, by using the inverse of the diagonalizing transformation, the eigenfunction envelopes are calculated. The dipole optical matrices are then calculated, and the absorption coefficient between states is also obtained for transitions involving all the bands. The model is applied in this chapter to calculation of the absorption coefficients of an intermediate band solar cell formed with quantum dots. The quantum efficiency is calculated and compared to measured values. The agreement is reasonable.

Keywords Solar cells · Quantum calculations · k·p methods · Quantum dots · Energy spectrum · Absorption coefficients

A. Luque and A.V. Mellor, *Photon Absorption Models in Nanostructured Semiconductor Solar Cells and Devices*, SpringerBriefs in Applied Sciences and Technology, DOI 10.1007/978-3-319-14538-9_3

3.1 Theoretical Background

The conditions allowing the use of the single band effective mass approximation are discussed in Sect. 2.1.3. Essentially, the condition is that $u_{v,k} \cong u_{v,0}$ for the band in consideration, or, as stated in Eq. (2.20), the single band approximation is valid if the only non-zero terms of the matrix elements of $E_{v,0} + U$ in basis of Bloch functions relates elements of a single band. In this section, the case in which this assumption fails is discussed.

The use of envolvent functions is the basis of the so-called k·p methods. The k·p methods introduced by Dresselhaus et al. [1] were extensively developed by Kane [2, 3] for calculations of semiconductor band structures. The topic is taught in several books [4, 5] including its application to nanostructured semiconductors. Even today, with the extensive use of modern computing facilities [6] the k·p methods are widely used and new books on the topic are still published [7–9]. However, they contain a background that is not often at the reach of device engineers. A book by Datta [10] (see Chap. 6) bridges this gap.

3.1.1 The Multiband Envelope Equations

Generalizing the single band envelope Eq. (2.15), we shall prove in the following sub-section that to obtain a solution of the exact Schrödinger Eq. (2.9) we first solve the multiband-envelope set of equations,

$$\sum_{\nu'} H_{\nu,\nu'}(-i\nabla)\Psi_{\nu'}(r) + {}^{\nu}U\Psi_{\nu}(r) = E\Psi_{\nu}(r) \tag{3.1}$$

$H_{\nu,\nu'}(-i\nabla)$ is obtained from $H_{\nu,\nu'}(k)$, calculated later in Eq. (3.9), by transforming $k \to -i\nabla$. ${}^{\nu}U$ contains a slow-varying potential added to the variation of the band edge caused by the nanostructure. There is a different potential function for each band (labeled by ν). The Hamiltonian eigenfunction is

$$\Xi(\boldsymbol{r}) \cong \sum_{\nu} u_{\nu,0}(\boldsymbol{r})\Psi_{\nu}(\boldsymbol{r}) \tag{3.2}$$

The preceding equations may refer to ordinary functions or to spinors. In the second case, each "function" $\Psi_{\nu}(\boldsymbol{r})$ is a two-component object and the square matrix elements are four-component objects. We can develop the thus-generated tensor products and consider that each band is a spin-up or a spin-down one with elements that are ordinary functions, leading to matrices with 4 times the number of elements of the matrix of spinors. This topic is further studied in Chap. 7.

3.1.2 Proof of the Development of the Hamiltonian Eigenfunction in a Series of Envelope Functions

We mainly follow reference [10] in this subsection. The set of functions $|0, v, \boldsymbol{k}\rangle = u_{v,0}(\boldsymbol{r})e^{i\boldsymbol{k}\cdot\boldsymbol{r}}/\sqrt{\Omega}$ form an orthonormal basis of the space of states. Due to the inhomogeneous nature of the nanostructured material, the functions $u_{v,0}(\mathbf{r})$ are not the same in the nanostructure and in the barrier material; however, we take them in a certain region and we call the corresponding material the *reference* semiconductor. Usually, it is either the QD material or the barrier material, depending the problem to be solved. The orthogonality of states of different $\boldsymbol{k}$ is assured by their plane-wave-like behavior. The orthogonality for different bands is assured by the orthogonality of the cell-periodic part of the Bloch functions, $u_{v,0}(\boldsymbol{r})$, for $\boldsymbol{k} = 0$ (which in this Γ-point is also the Bloch function, with acronym GBF), as shown in Eq. (2.13). We can develop the wavefunction as

$$\Xi(\mathbf{r}) \equiv \sum_{v,\mathbf{k}} \psi_{v,\mathbf{k}}|0, v, \mathbf{k}\rangle; \quad |0, v, \mathbf{k}\rangle \equiv u_{v,0}(\mathbf{r})e^{i\mathbf{k}\cdot\mathbf{r}}/\sqrt{\Omega} = u_{v,0}(\mathbf{r})|\mathbf{k}\rangle \tag{3.3}$$

where $|\mathbf{k}\rangle$ is the ordinary plane wave already defined in Eq. (2.21). However, the basis-functions $|0, v, \boldsymbol{k}\rangle$ are not eigenfunctions of the lattice Hamiltonian of Eq. (2. 10), unless if $\boldsymbol{k} = 0$. In effect,

$$\begin{aligned} H|0, v, \boldsymbol{k}\rangle &= -\frac{\hbar^2}{2m_0}\left(\frac{e^{i\mathbf{k}\cdot\mathbf{r}}\nabla^2 u_{v,0}(\boldsymbol{r}) + 2i\boldsymbol{k}e^{i\mathbf{k}\cdot\mathbf{r}}\nabla u_{v,0}(\boldsymbol{r}) - k^2 u_{v,0}(\boldsymbol{r})e^{i\boldsymbol{k}\cdot\boldsymbol{r}}}{\sqrt{\Omega}}\right) + U_L(\boldsymbol{r})\frac{u_{v,0}(\boldsymbol{r})e^{i\boldsymbol{k}\cdot\boldsymbol{r}}}{\sqrt{\Omega}} \\ &= \left(E_{v,0} + \frac{\hbar^2 k^2}{2m_0}\right)\left(\frac{u_{v,0}(\boldsymbol{r})e^{i\boldsymbol{k}\cdot\mathbf{r}}}{\sqrt{\Omega}}\right) - \frac{i\hbar^2}{m_0}\left(\frac{e^{i\boldsymbol{k}\cdot\boldsymbol{r}}\boldsymbol{k}\nabla u_{v,0}(\boldsymbol{r})}{\sqrt{\Omega}}\right) \neq E_{v,\boldsymbol{k}}|0, v, \boldsymbol{k}\rangle \end{aligned} \tag{3.4}$$

As already said, $u_{v,0}$ has been chosen to correspond to a homogeneous reference semiconductor, either the one corresponding to the barrier or to the nanostructure. For $\boldsymbol{k} = 0$, the equality is fulfilled with the use of the dispersion equation of Eq. (2.11) containing the definition of $E_{\boldsymbol{k},v}$.

H is the reference semiconductor's Hamiltonian and refers to a homogeneous material. The effective mass will be considered that of the reference material, but we cannot ignore the variation of the band edges. In this section, the band edge will be expressed as $E_{v,0} + \Delta E_{v,0}$, the later part being the variation with respect to the edge of the reference semiconductor. Thus, the development of the TISE Eq. (2.9) in this basis (as found by introducing Eq. (3.3) in the TISE Eq. (2.9), with primes in the running letters, and left-multiplying by $\langle 0, v, \boldsymbol{k}|$) leads to the set of equations.

$$\sum_{v',\boldsymbol{k}'} \left(\langle 0,v,\boldsymbol{k}|H - \Delta E_{v,0}|0,v',\boldsymbol{k}'\rangle + \langle 0,v,\boldsymbol{k}|\Delta E_{v,0} + U|0,v',\boldsymbol{k}'\rangle \right) \psi_{v',\boldsymbol{k}'} = E\psi_{v,\boldsymbol{k}} \tag{3.5}$$

Note that we have subtracted and added $\Delta E_{v,0}$ (with the band index of the bra vector) to the two elements of matrix. The transformation $-i\nabla \rightarrow \boldsymbol{k}$ is strictly valid when applied to a function independent of $\boldsymbol{r}$. We have rearranged the terms to remove the $\boldsymbol{r}$-dependent part from the first term when the dispersion function of Eq. (2.11) is used, adding it to the second term. To simplify the notation, this equation will simply be written as,

$$\sum_{v',\boldsymbol{k}'} \left(\langle 0,v,\boldsymbol{k}|H(\boldsymbol{k}')|0,v',\boldsymbol{k}'\rangle + \langle 0,v,\boldsymbol{k}|^{v}U(r)|0,v',\boldsymbol{k}'\rangle \right) \psi_{v',\boldsymbol{k}'} = E\psi_{v,\boldsymbol{k}} \tag{3.6}$$

where H is the reference Hamiltonian with no nanostructure (and is the expression obtained of the equality in Eq. (3.4) but the $\boldsymbol{r}$-variable term is now called ${}^{v}U = E_{v,0} + U$ and bears a band index because it is band dependent.

The space dependent term ${}^{v}U$ is a slowly variable function, therefore, by application of the integral factorization rule of Eq. (2.1),

$$\langle 0,v,\boldsymbol{k}|^{v}U(\boldsymbol{r})|0,v',\boldsymbol{k}'\rangle = \delta_{v,v'}\langle \boldsymbol{k}|^{v}U(\boldsymbol{r})|\boldsymbol{k}'\rangle \equiv \delta_{v,v'}{}^{v}U_{\boldsymbol{k},\boldsymbol{k}'} \tag{3.7}$$

Furthermore, looking at Eq. (3.6) and taking into account that $\boldsymbol{k}'$ are treated as parameters in the matrix element (whose integration is on $\boldsymbol{r}$) and applying the orthogonality properties of plane waves,

$$\langle 0,v,\boldsymbol{k}|H(\boldsymbol{k}')|0,v',\boldsymbol{k}'\rangle = H_{v,v'}(\boldsymbol{k})\delta_{\boldsymbol{k},\boldsymbol{k}'} \tag{3.8}$$

with

$$\begin{aligned} &\forall v = v'; \quad H_{v,v}(\boldsymbol{k}) = E_{v,0} + \frac{\hbar^2 k^2}{2m_0} \\ &\forall v \neq v'; \quad H_{v,v'}(\boldsymbol{k}) = \frac{\hbar\boldsymbol{k}\cdot P_{v,v'}}{m_0}; \quad P_{v,v'} = -i\hbar\langle u_{v,0} \mid \nabla u_{v',0}\rangle \end{aligned} \tag{3.9}$$

This is why this model is called the $\boldsymbol{k}\cdot\boldsymbol{p}$ model. Remember that $E_{v,0}$ relates to the reference semiconductor and is therefore a constant, different for each band. The $\boldsymbol{P}_{v,v'}$ elements are sometimes called Kane matrix elements.

The latter elements of matrix are obtained by left-multiplying Eq. (3.4) by $\langle 0,v,\boldsymbol{k}|$. The integral factorization rule is also applied here. The term $\boldsymbol{P}_{v,v}$ is zero. This can be easily proven in case of cubic symmetry. In effect, due to the periodicity of $u_{v,0}$,

$$
\begin{aligned}
P_{v,v,z} &= \frac{-i\hbar}{2\Omega_{cell}} \int_{-L_{cell}/2}^{L_{cell}/2} dx \int_{-L_{cell}/2}^{L_{cell}/2} dy \int_{-L_{cell}/2}^{L_{cell}/2} \frac{d\left|u_{v,0}\right|^2}{dz} dz \\
&= \frac{-i\hbar}{2\Omega_{cell}} \int_{-L_{cell}/2}^{L_{cell}/2} dx \int_{-L_{cell}/2}^{L_{cell}/2} dy \left|u_{v,0}\right|^2 \Big|_{-L_{cell}/2}^{L_{cell}/2} = 0
\end{aligned} \tag{3.10}
$$

and the same for the other $P_{v,v}$ components.

Using Eqs. (3.7) and (3.8) we can rewrite Eq. (3.6) as

$$\sum_{v'} H_{v,v'}(\boldsymbol{k})\psi_{v',\boldsymbol{k}} + \sum_{\boldsymbol{k}'} {}^{v}U_{\boldsymbol{k},\boldsymbol{k}'}\psi_{v,\boldsymbol{k}'} = E\psi_{v,\boldsymbol{k}} \tag{3.11}$$

Let us now develop the envolvent functions for each band as sums of plane waves

$$\Psi_v(\boldsymbol{r}) = \sum_{\boldsymbol{k}} \psi_{v,\boldsymbol{k}}|\boldsymbol{k}\rangle \tag{3.12}$$

Substituting this definition into Eq. (3.1) to develop this equation in a basis of plane waves, after putting primes in the $\boldsymbol{k}$ sum index, we obtain

$$\sum_{v',\boldsymbol{k}'} H_{v,v'}(\boldsymbol{k}')\psi_{v,\boldsymbol{k}'}|\boldsymbol{k}'\rangle + \sum_{\boldsymbol{k}} {}^{v}U\psi_{v,\boldsymbol{k}'}|\boldsymbol{k}'\rangle = E\sum_{\boldsymbol{k}} \psi_{v,\boldsymbol{k}'}|\boldsymbol{k}'\rangle \tag{3.13}$$

Left-multiplying by $\langle\boldsymbol{k}|$,

$$\sum_{v'} H_{v,v'}(\boldsymbol{k})\psi_{v',\boldsymbol{k}} + \sum_{\boldsymbol{k}'} \langle\boldsymbol{k}|^{v}U|\boldsymbol{k}'\rangle\psi_{v,\boldsymbol{k}'} = E\psi_{v,\boldsymbol{k}} \tag{3.14}$$

which is the same as Eq. (3.11). This means that the same coefficients are valid for the envelope function and for the TISE solution. This proves that

$$\Xi(\boldsymbol{r}) = \sum_{v,\boldsymbol{k}} u_{v,0}(\boldsymbol{r})\psi_{v,\boldsymbol{k}}|\boldsymbol{k}\rangle = \sum_{v} u_{v,0}(\boldsymbol{r})\Psi_v(\boldsymbol{r}) \tag{3.15}$$

q.e.d.

3.1.3 The Hamiltonian Matrix in Zincblende Materials

InAs, GaAs and many other semiconductors in use in nanotechnology crystallize as the zincblende. In this case, the orbital part of the GBF wavefunctions is represented by $|X\rangle, |Y\rangle, |Z\rangle, |S\rangle$ (the X-, Y-, Z- and S-GBFs) with the symmetry of $x(x^2+y^2+z^2)$, $y(x^2+y^2+z^2)$, $z(x^2+y^2+z^2)$ and $(x^2+y^2+z^2)$ respectively [4, 10]. In the four band model, the spin coordinates are neglected. In other words, each orbital wavefunction is doubly degenerate with spin 1/2 or −1/2. We adopt this simplification because we are looking for the simplest Hamiltonian able to construe the semiconductor energy bands. The envelope functions that multiply the X-, Y-, Z- and S-GBFs are called the X-, Y-, Z-, and S-envelopes. Furthermore, due to this specific symmetry,

$$\langle S| - i\hbar\frac{\partial}{\partial x}|X\rangle = \langle S| - i\hbar\frac{\partial}{\partial y}|Y\rangle = \langle S| - i\hbar\frac{\partial}{\partial z}|Z\rangle = P_0 \qquad (3.16)$$

with all the remaining terms and components of $\boldsymbol{P}_{\nu,\nu'}$ being zero. The four-band $H_{\nu,\nu'}(\boldsymbol{k})$ terms for zincblende semiconductors are collected in Fig. 3.1 ("Hamiltonian Matrix" panel). This simplified matrix does not consider the spin orbit interaction [10], nor the strain effects due to the inclusion of the QDs, though these are both important. We may call it zero order Hamiltonian matrix of (H_0).

The eigenvalues, also written in the same figure ("Eigenvalues" panel), show four values (each one with two spin states) for every $\boldsymbol{K}$ ($\boldsymbol{k}$ normalized to $1/d$). From top to bottom, the first eigenvalue corresponds to the CB, the second to the light hole VB, and the third and fourth to a doubly degenerate VB state. The doubly degenerate VB eigenvalues present a positive effective mass which is known to be impossible (these bands would fill the band gap). Thus, the assumption of no spin-orbit interaction and neglecting the effect of the lattice strain are not acceptable within the four band k·p model.

The eigenvectors corresponding to the mentioned (non-acceptable) eigenvalues are also represented in Fig. 3.1, in the "Eigenvectors" panel (each eigenvector is a given as a row adjacent to the corresponding eigenvalue). The doubly degenerate eigenvalue defines a two-dimensional space of vectors all of which are normal to the other two one-dimensional eigenvectors. In this two-dimensional space, the choice of the basis is arbitrary. We have taken the choice in [11] that differs from the one in [12].

Notice that the four eigenvectors are mutually orthogonal. Each eigenvector, once normalized, forms a row of a unitary matrix (T) such that $(H_{0,d}) = (T)\,(H_0)\,(T^{+})$ forms a diagonal matrix whose elements are the eigenvalues of (H_0). The superscript + means the Hermitical conjugate (the complex conjugated of the transposed) matrix.

In [11], the VB top was taken as the energy origin. In our Fig. 3.1, no origin has been selected; instead the bottom of the CB (E_C) and the top of the VB (E_V) are used explicitly. Actually this change of origin is reflected in the eigenvalues but not in the eigenvectors; therefore (T) remains unchanged. Actually, changing the origin is the equivalent to subtracting from (H_0) the origin $E_{or}(I)$, (I) being the unity matrix. Then, $(T)[(H_0) - E_{or}(I)](T^{+}) = (H_{0,d}) - E_{or}(I)$. By setting $E_{or} = E_V$ we recover the expressions in [11]. The reason we prefer not to set $E_V = 0$ is because, in

Hamiltonian Matrix

$H_{v,v'}(\boldsymbol{k}) = \frac{1}{\Omega_{cell}} \int u^*_{v,0} e^{-i\boldsymbol{k}\boldsymbol{r}} H u_{v',0} e^{i\boldsymbol{k}\boldsymbol{r}} d^3r$	$\vert X\rangle$	$\vert Y\rangle$	$\vert Z\rangle$	$\vert S\rangle$
$\langle X\vert$	$E_V + AK$	0	0	BK_x
$\langle Y\vert$	0	$E_V + AK$	0	BK_y
$\langle Z\vert$	0	0	$E_V + AK$	BK_z
$\langle S\vert$	BK_x	BK_y	BK_z	$E_C + AK$

Eigenvalues	**Eigenvectors**				**Proposed Eigenvalues**
$E_{cb}(K) = E_v + E_g/2 + AK^2 + \sqrt{(E_g/2)^2 + B^2K^2}$	K_x	K_y	K_z	$\frac{E_g + \sqrt{E_g^2 + 4B^2K^2}}{2B}$	$E_{cb}(K) \cong E_C + \frac{\hbar^2K^2}{2m_{cb}d^2}$
$E_{lh}(K) = E_v + E_g/2 + AK^2 - \sqrt{(E_g/2)^2 + B^2K^2}$	K_x	K_y	K_z	$\frac{E_g - \sqrt{E_g^2 + 4B^2K^2}}{2B}$	$E_{lh}(K) \cong E_V - \frac{\hbar^2K^2}{4m_{lh}d^2}$
$E_{v1}(K) = E_v + AK^2$	$-K_y$	K_x	0	0	$E_{hh}(K) \cong E_V - \frac{\hbar^2K^2}{2m_{hh}d^2}$
$E_{v2}(K) = E_v + AK^2$	K_xK_z	K_yK_z	$-(K_x^2 + K_y^2)$	0	$E_{so}(K) \cong E_{SO} - \frac{\hbar^2K^2}{2m_{so}d^2}$

$E_g = E_C - E_V$; $A = \frac{\hbar^2}{2m_0d^2}$; $B = \frac{\hbar P_0}{m_0 d}$; $\boldsymbol{K} = \boldsymbol{k}d$; $P_0 = \langle S|p_x|X\rangle = \langle S|p_y|X\rangle = \langle S|p_z|X\rangle$

The origin of energies is at the Γ point of the valence band; d: normalizing arbitrary length

Fig. 3.1 Hamiltonian matrix, eigenvalues and eigenvectors when the spin-orbit coupling is neglected and proposed eigenvalues. Reproduced with permission. © 2012, Elsevier [11]

heterogeneous materials, E_V is not a constant throughout the whole material, and indeed nor is E_C.

Assuming that B^2K^2 in Fig. 3.1 is small with respect to E_g^2, we can develop E_{cb} (in the "Eigenvalues" panel) in series, leading to

$$E_{cb}(K) - E_V \cong E_g + AK^2 + \frac{B^2K^2}{E_g} \equiv E_g + \frac{\hbar^2K^2/d^2}{2m_{cb}} \tag{3.17}$$

the last equality being a definition of the effective mass within this model. Experimentally obtained values for m_{cb} are tabulated in the literature for many bulk semiconductors. This allows P_0 to be obtained from experiments, since, from Eq. (3.17),

$$
\begin{aligned}
B^2 &= \frac{\hbar^2 E_g}{2m_0 d^2}\left(\frac{m_0}{m_{cb}} - 1\right) = AE_g\left(\frac{m_0}{m_{cb}} - 1\right) \\
P_0 &= \sqrt{\frac{E_g}{2} m_0 \left(\frac{m_0}{m^*_{cb}} - 1\right)}
\end{aligned}
\tag{3.18}
$$

The effective mass for light holes may be deduced form the series development of E_{lh} in the "Eigenvalues" panel, but it might not agree very well with the actual experimental value. The value of P_0 and of the effective masses are usually derived from the Luttinger parameters [10] (p. 167) that are also tabulated for most semiconductors and ultimately are experimental parameters. Chapter 7 deals with this subject more in depth.

3.1.4 The Fourier Transforms and Plane Wave Developments

The use of some kind of Fourier transform is a practical way of obtaining a plane wave development. In some cases the wavefunctions $\Phi(\boldsymbol{r}) = \xi(x)\varphi(y)\zeta(z)$ are the product of three one-dimensional functions. In this case, the Fourier transform is the product of three one dimensional Fourier transforms. In general the three dimensional Fourier transform is obtained by the successive application of three one-dimensional transforms. In this work the Discrete Fourier Transform (DFT) and the Fourier Integral (FI) are used for bound and extended wavefunctions respectively.

One of these one-dimensional functions developed in plane waves is

$$
\zeta_v(z) = \sum_{k} \varphi_z(k_z) \frac{e^{ik_z z}}{\sqrt{L_z}}
\tag{3.19}
$$

where L_z is the length of the z-space, which is usually large. With this definition of the plane wave development, the DFT and its inverse are respectively,

$$
\begin{aligned}
\sqrt{\frac{N}{L_z}} \varphi_{z,\kappa_z} &= \mathscr{F}\left[\{\zeta_{n_z}\}\right]_{\kappa_z} \equiv \frac{1}{\sqrt{N}} \sum_{n_z=-(N-1)/2}^{(N-1)/2} \zeta_{n_z} e^{-2\pi i \kappa_z n_z / N} \\
\sqrt{\frac{L_z}{N}} \zeta_{n_z} &= \mathscr{F}^{-1}\left[\{\varphi_{z,\kappa_z}\}\right]_{n_z} = \frac{1}{\sqrt{N}} \sum_{\kappa_z=-(N-1)/2}^{(N-1)/2} \varphi_{z,\kappa_z} e^{2\pi i \kappa_z n_z / N}
\end{aligned}
\tag{3.20}
$$

where N is the number of calculation points, which is always odd. $\equiv$ indicates a definition. To obtain the DFT, $\zeta(z)$ is calculated at a grid of points $\zeta_{n_z} = \zeta(n_z L_z / N)$. $\varphi_{z,\kappa_z} = \varphi(2\pi\kappa_z / L_z)$ is calculated at the permitted points of k_z, with κ_z being an integer, The inverse DFT (IDFT) of the set $\{\varphi_{z,\kappa_z}\}$ gives a set of numbers proportional to $\{\zeta_{n_z}\}$ (in this chapter, *braces* or curled brackets enclose sets of values).

For extended functions, such as $\xi(x)$, whose development in plane waves is

$$\xi(x) = \int \varphi_x(k_x) \frac{e^{ik_x x}}{\sqrt{L_x}} dk_x \tag{3.21}$$

it is better to use the FI,

$$\begin{aligned}
\sqrt{\frac{2\pi}{L_x}}\varphi(k_x) &= \mathscr{F}[\xi(x)](k_x) \equiv \frac{1}{\sqrt{2\pi}} \int_{-\infty}^{\infty} \xi(x) e^{-ik_x x} dx \\
\sqrt{\frac{L_x}{2\pi}}\xi(x) &= \mathscr{F}^{-1}[\varphi(k_x)](x) = \frac{1}{\sqrt{2\pi}} \int_{-\infty}^{\infty} \varphi(k_x) e^{ik_x x} dk_x
\end{aligned} \tag{3.22}$$

In both sets of formulas, the Fourier Transform is denoted with a script F and its argument is between (squared) brackets. This argument is a function that shows explicitly its variable for the FI and its inverse (IFI). The brackets are followed by the transform variable between parentheses (or round brackets). In contrast, the argument for the DFT and its inverse is a set of values between *braces* (or curly brackets), which shows the subindex of the set for each element, the transform presents a subindex outside the brackets for the transformed variable.

3.2 The Empiric k·p Hamiltonian Approximation

The Empiric k·p Hamiltonian (EKPH) approximation utilizes the simplest Hamiltonian able to explain the existence of a forbidden band in a semiconductor. It is a four band k·p approximation in which every state is assumed to be doubly-spin degenerate. It is based on the Hamiltonian appearing in Fig. 3.1.

One basic assumption of the EKPH approximation is that the eigenvalues ("Proposed Eigenvalues" panel) follow parabolic shapes with the experimental effective masses. The fact that the effective masses are positive for the conduction band and negative for the valence band makes the new Hamiltonian that these eigenvalues are creating experimentally consistent. With these eigenvalues, we can build a new diagonal Hamiltonian ($H_{EKP,d}$) that takes implicitly into account the spin-orbit coupling and the crystal strain. The fact that the bands are taken to be isotropic erases any memory of the crystalline structure of the materials involved so that the only relevant symmetry is that of the nanostructure (although the GBFs keep their own symmetry). This diagonal Hamiltonian is enough to obtain all the eigenvalues of the nanostructure as we shall see in Sect. 3.2.1.

A second basic assumption of the EKPH model is the choice of the diagonalization matrix that restores the Hamiltonian matrix to the initial basis functions, expressed in the definition of the matrix elements appearing in Fig. 3.1. We assume, in a first instance, that this matrix is the matrix (T), already defined, which is based on the eigenvectors of the "Eigenvectors" panel (once normalized), so that $(H_{EKP}) = (T^{+})(H_{EKP,d})(T)$. In summary, this consists in supposing that (H_0) and (H_{EKP}) have the same eigenvectors but different eigenvalues. This is the same assumption taken in [11, 12]. However, it is to be understood that the eigenvectors appearing in Fig. 3.1 are not unambiguous because any orthogonal couple of vectors of the subspace spanned by the double degenerate eigenvalue may be a possible choice of the basis eigenvectors.

All these basis vectors may be obtained by applying to (T) the unitary transformation corresponding to a Givens rotation matrix:

$$(R) = \begin{pmatrix} 1 & 0 & 0 & 0 \\ 0 & 1 & 0 & 0 \\ 0 & 0 & \sqrt{1-\gamma^2} & \gamma \\ 0 & 0 & -\gamma & \sqrt{1-\gamma^2} \end{pmatrix} \tag{3.23}$$

which can be interpreted as a rotation of an angle $\arcsin(\gamma)$ around a (two-dimensional) axis that leaves invariant the subspace corresponding to the non-degenerate eigenvalues of (H).

The resulting diagonalization matrix (T_γ) is decomposed for presentation purposes into two summands $(T_\gamma) = (T_{\gamma A}) + (T_{\gamma B})$, which appear in Table 3.1. It is to be stressed that $(T_{\gamma A})$ and $(T_{\gamma B})$ are not diagonalization matrices: they are not unitary. Note that when $\gamma = 0$, $(T_{\gamma B})$ vanishes and $(T_{\gamma A})$ becomes equal to the matrix (T_γ) appearing in Fig. 3.1, once normalized with the row norms of this matrix, N_1, N_2, N_3, and N_4. Some terms disappear in both matrices for $\gamma = 1$.

Table 3.1 Summand matrices $(T_{\gamma A})$ and $(T_{\gamma B})$ into which the rotated transformation matrix (T_γ) is decomposed

$$(T_{\gamma A}) = \begin{pmatrix} K_x/N_1 & K_y/N_1 & K_z\sqrt{1-\gamma^2}/N_1 & \left(E_G + \sqrt{E_G^2 + 4B^2(K_x^2+K_y^2+K_z^2)}\right)\sqrt{1-\gamma^2}/2BN_1 \\ K_x/N_2 & K_y/N_2 & K_x\sqrt{1-\gamma^2}/N_2 & \left(E_G - \sqrt{E_G^2 + 4B^2(K_x^2+K_y^2+K_z^2)}\right)\sqrt{1-\gamma^2}/2BN_2 \\ -K_y\sqrt{1-\gamma^2}/N_3 & K_x\sqrt{1-\gamma^2}/N_3 & 0 & 0 \\ K_xK_z\sqrt{1-\gamma^2}/N_4 & K_yK_z\sqrt{1-\gamma^2}/N_4 & -(K_x^2+K_y^2)(1-\gamma^2)/N_4 & 0 \end{pmatrix}$$

$$(T_{\gamma B}) = \gamma \begin{pmatrix} 0 & 0 & \left(E_G + \sqrt{E_G^2 + 4B^2(K_x^2+K_y^2+K_z^2)}\right)/2BN_1 & -K_z/N_1 \\ 0 & 0 & \left(E_G - \sqrt{E_G^2 + 4B^2(K_x^2+K_y^2+K_z^2)}\right)/2BN_2 & -K_z/N_2 \\ K_xK_z/N_4 & K_yK_z/N_4 & -(K_x^2+K_y^2)(1-\gamma^2)/N_4 & (K_x^2+K_y^2)\gamma/N_4 \\ K_5/N_3 & -K_x/N_3 & 0 & -(K_x^2+K_y^2)\sqrt{1-\gamma^2}/N_3 \end{pmatrix}$$

$$N_1 = \sqrt{8B^2(K_x^2+K_y^2+K_z^2) + 2E_g\left(E_g + \sqrt{E_g^2 + 4B^2(K_x^2+K_y^2+K_z^2)}\right)} \Big/ 2B$$

$$N_2 = \sqrt{8B^2(K_x^2+K_y^2+K_z^2) + 2E_g\left(E_g - \sqrt{E_g^2 + 4B^2(K_x^2+K_y^2+K_z^2)}\right)} \Big/ 2B$$

$$N_3 = \sqrt{(K_x^2+K_y^2)}; \quad N_4 = \sqrt{(K_x^2+K_y^2)(K_x^2+K_y^2+k_z^2)};$$

3.2.1 The Diagonalized Hamiltonian

Applying the multiband envelope equations of Eq. (3.2) to the diagonalized Hamiltonian for an eigenfunction of energy corresponding to the conduction band (cb), the envelope equation in a region without charge of space is

$$-\frac{\hbar^2}{2m_{cb}}\nabla^2\Phi + E_C(\mathbf{r})\Phi = E\Phi \tag{3.24}$$

where E_C is variable with the position. To solve this part of the problem, the GaAs CB bottom can be taken as the energy origin (at both sides of the equality). Thus, in the InAs QD there is an E_C well that we may consider squared (in energy) and box-shaped (in space). In this way, the work developed for the effective mass one-band model may be applied to this case. However, for presentation, the VB top may be used as the energy origin, as it is done in Fig. 3.4.

The effective mass is also different in the GaAs barrier material than in the InAs QD. However, in our model we must take it as a constant. We may choose its value to be that for the InAs, for example, if the wavefunctions are mainly localized there (bound wavefunctions). A modification of the effective mass depending on the region is discussed in Sect. 2.1.5 but is not considered generally in this book.

The use of Eq. (3.24) allows us to calculate the diagonalized Hamiltonian eigenvalues and eigenfunctions.

For an eigenfunction corresponding to one of the VBs, e.g. the light hole (lh) valence band, the corresponding Schrödinger equation is, according to Fig. 3.1,

$$-\frac{\hbar^2}{2m_{lh}}\nabla^2\Phi - E_v(\mathbf{r})\Phi = -E\Phi \tag{3.25}$$

Note that this equation has been changed of sign entirely. This is to compensate the negative character of the effective mass. Thanks to this, a pedestal in the VB top, as occurs in Type I QDs, is equivalent to a potential well in the CB. Again, here the potential origin may be located (at least for presentation) at the GaAs VB top, but the sign of the energy levels and the different energy origin in each calculation must be accounted for when a global plot is presented.

The same may be applied to all rest of the VBs: heavy holes (hh) and split-off (so). Notice that, in this work, VB and CB in capital letters refer to the range of energies above and below the band edges in the barrier material (GaAs) whereas *hh*, *lh*, *so* and *cb* refer to the states created by the QDs and their energy can also be located within the barrier material bandgap.

As in Chap. 2, the QD shape is simplified to a squat parallelepiped or box as represented in Fig. 3.2. Furthermore, for all the bands, the offsets form the confining potential; they are assumed to be square, that is, in any direction traversing the QD box, the band edge outside the QD has the position of the band edge in the homogeneous host material and inside the QD has the band edge of the QD

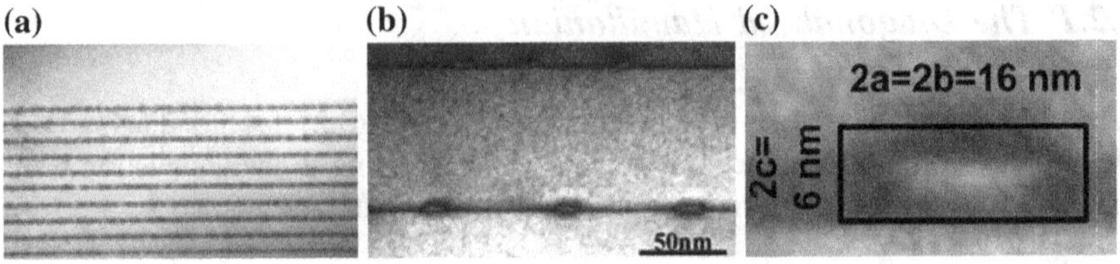

Fig. 3.2 Transmission electron microscope picture of the multiple layers forming the intermediate band region of an intermediate band solar cell. From **a** to **c**, increasing magnification. **c** The quantum dot box modeling. The continuous line mainly visible in **b** is the so called wetting layer (WL), a continuous layer of QD material formed naturally in the Stranski-Krastanov growth method

material. This assumption is somewhat unrealistic because the QD is strongly affected by the strain produced by the QDs which differ with the position inside and outside the QD, but it is easy to handle and agrees with experimental results as shown e.g. in Refs. [11–13]. This subject is further studied in Chap. 7.

To appraise the degree of approximation of the square offset potential, we present in Fig. 3.3 an example of published calculations by Popescu et al. [14] with a lens shape and a diameter similar that of the QDs in this paper, but a height which is about half. The band offset is calculated using an elastic model of the QD/host strain [15–17]. In (b) and (c) the InAs/GaAs is the same structure of the SOTA prototype modeled along this chapter. A strong reduction of the CB offset is observed due to the strain, together with a substantial increase of the VB pedestal.

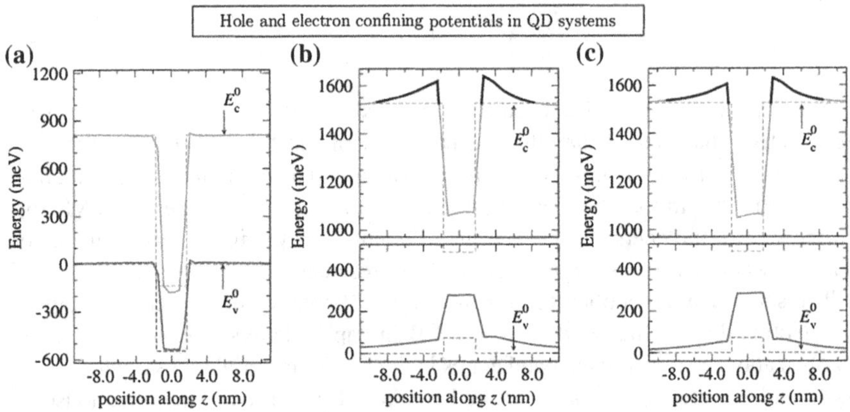

Fig. 3.3 Line plots of the hole (magenta/grey) and electron (cyan/light grey) strain-modified confining potentials for various QD systems: **a** unstrained InAs/GaSb without WL; **b** strained InAs/GaAs without WL; **c** strained InAs/GaAs with a 4 monolayers thick WL. The unstrained band edges of the corresponding matrix and *dot* materials are shown with *dashed lines*, and labeled as *Ec*0 and *Ev*0, respectively. For all the plots, the confining potentials are shown along a line going through the *dot center* parallel to the *z* axis. Energy zero is the VB maximum of each of the matrices. Reproduced with permission from [14]. © 2009 APS

The InAs bandgap passes from 0.418 to about 0.737 eV for the InAs strained in the QD. (c) shows that no big difference is produced by the WL. In (a) the InAs QD in a GaSb host is shown. The VB pedestal is changed into a well. This forms the so-called Type II QD, as opposed to the Type I QDs of cases (b) and (c). Type II QDs have interesting properties for IBSCs [18].

For the squat box shape, the separation of variables method, described in Sect. 2. 2.1 can be used [12]. The eigenfunctions are labeled by three quantum numbers (QNs) each one corresponding to the one-dimensional solution chosen, from less to more energetic; for example, the eigenfunction can be labeled *cb*(1,2,1), etc. Odd QNs refer to even one-dimensional solutions and vice versa. This has already been widely discussed for the one-band approximation, but, in this case, we add the prefix *cb* to mean that we are referring to one of the *cb* states. Alternatively we might write $|cb121\rangle$.

In Fig. 3.4, we reproduce the energy levels corresponding to the SOTA prototype QDs. As a consequence of the large effective mass, there are many *hh* eigenvalues. A density of states rather than the energy levels is presented in this case.

A better approximation is the first order approximation, which is explained in Sect. 2.3.2.1. Figure 3.5 shows, for the SOTA prototype, the difference between using the pure separation of variables and the 1st order correction of it in the *cb*, *lh* and *hh* bands. The corrections affecting the IB states are negligible but the virtual bound states within the CB are to be brought down substantially, in particular for the highest energies. The VB states separable calculation is more accurate.

Figure 3.6 represents the spread of the wavefunctions for the (1,2,1) QNs corresponding to the *cb*, the *lh* and the *hh* band. Notice that the high effective mass of the *hhs* makes the electron stay strongly confined within the QD. Contrarily, the *lh* states are less strongly confined because of their light effective mass and the weak confining potential. The *cb* electron is more confined due to the stronger potential.

3.2.2 The Envelope Functions

In the k·p method, the envelope functions result from a plane wave development as indicated in Eq. (3.12). In practice, the assumption of a finite number of plane waves results in the determination of the wavefunction at a finite number of nodes of a calculation cubic lattice. The rest of the points, if needed, are to be interpolated. Mathematically speaking, the coefficients $\psi_v(2\pi\boldsymbol{\kappa}/L)$ of the development in Eq. (3.12), that we shall also call $\psi_{v,\boldsymbol{\kappa}}$, are the Discrete Fourier Transform (DFT) of the values of $\Delta\Omega^{1/2}\ \Psi_v(\boldsymbol{r})$ calculated at the lattice nodes, where $\Delta\Omega = (L/N)^3 = l^3$ is the volume element associated to a calculation node and $\Omega = \Delta\Omega \times N^3 = L^3$. The values of $\Delta\Omega^{1/2}\Psi_v(\boldsymbol{n}L/N)$ at these nodes ($\boldsymbol{n}$, like $\boldsymbol{\kappa}$, are triplets of integers) are called $\Psi_{v,\boldsymbol{n}}$ and they are the inverse transform (IDFT) of $\psi_{v,\boldsymbol{\kappa}}$.

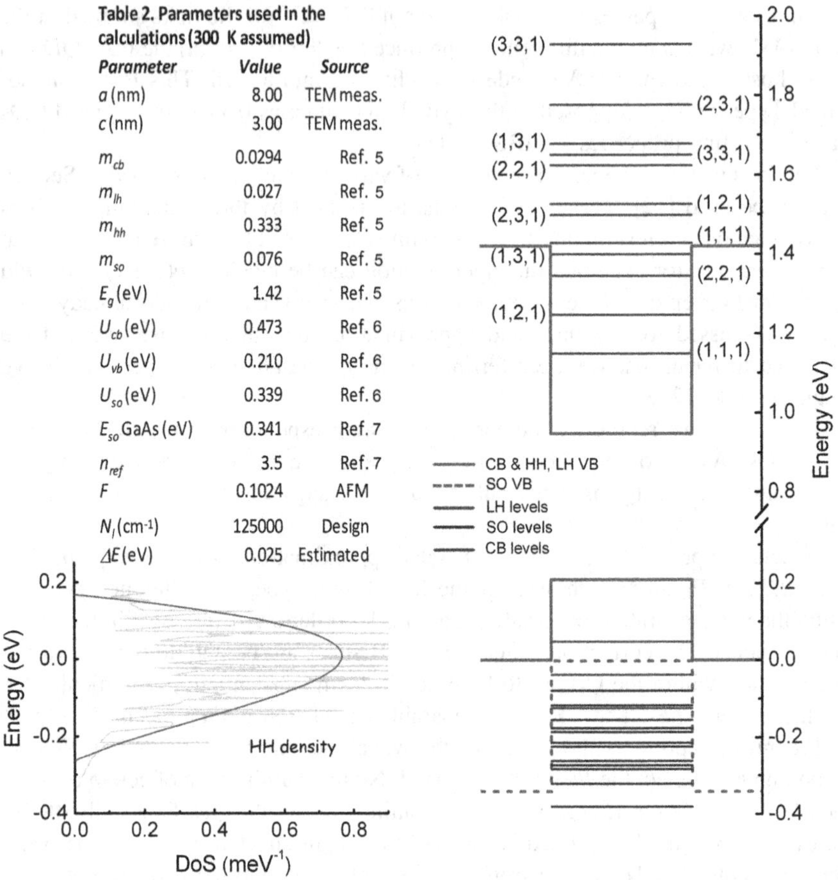

Table 2. Parameters used in the calculations (300 K assumed)

Parameter	*Value*	*Source*
a (nm)	8.00	TEM meas.
c (nm)	3.00	TEM meas.
m_{cb}	0.0294	Ref. 5
m_{lh}	0.027	Ref. 5
m_{hh}	0.333	Ref. 5
m_{so}	0.076	Ref. 5
E_g (eV)	1.42	Ref. 5
U_{cb} (eV)	0.473	Ref. 6
U_{vb} (eV)	0.210	Ref. 6
U_{so} (eV)	0.339	Ref. 6
E_{so} GaAs (eV)	0.341	Ref. 7
n_{ref}	3.5	Ref. 7
F	0.1024	AFM
N_I (cm^{-1})	125000	Design
ΔE (eV)	0.025	Estimated

Fig. 3.4 Energy levels in the QD and data for their calculation. There are 32, spin-degenerate, derived for the *cb*, 4 for the *lh* band, 18 for the *so* band and 147 for the *hh* band. The latter are represented by their Density of States (DoS) adjusted by a polynomial. 13 *cb* states, the (1,1,1) state in the *so* band, all the *lh* states and many *hh* states have their energies in the range required to contribute to the sub-bandgap photocurrent. Data from measurements and Refs. [14, 19, 20]. Reproduced with permission. © 2011, Elsevier [12]

The same nomenclature can be associated to the diagonalized eigenfunctions. By using only the same discrete set of points $\Phi_{u,\mathbf{n}} \equiv \Delta\Omega^{1/2}\Phi_u(\mathbf{n}L/N)$ whose DFT are $\phi_{u,\boldsymbol{\kappa}} = \varphi_u(2\pi\boldsymbol{\kappa}/L)$, with u referring to the *cb*, *lh*, *hh*, or *so* band, the following relationship can be written,

$$\phi_{u,\boldsymbol{\kappa}} = \frac{1}{N^{3/2}}\sum_{\mathbf{n}} \Phi_{u,\mathbf{n}} e^{-2\pi i\boldsymbol{\kappa}\cdot\mathbf{n}/N} \tag{3.26}$$

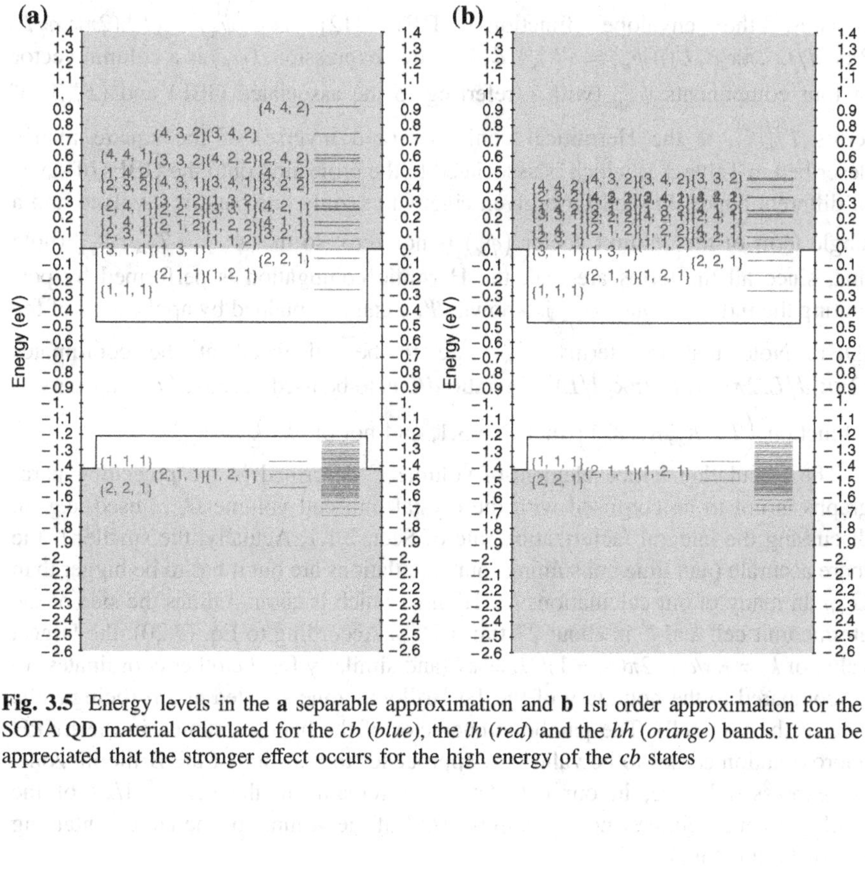

Fig. 3.5 Energy levels in the **a** separable approximation and **b** 1st order approximation for the SOTA QD material calculated for the *cb* (*blue*), the *lh* (*red*) and the *hh* (*orange*) bands. It can be appreciated that the stronger effect occurs for the high energy of the *cb* states

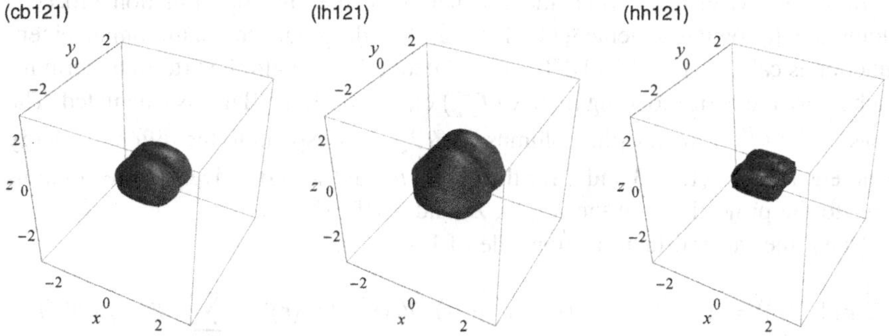

Fig. 3.6 Equal-probability-density contours corresponding to 0.1 % of their maximum value for diagonal Hamiltonian eigenfunctions of QNs (1,2,1) for different bands. The *green box* represents the QD. Space coordinates are in units of d (d = 10 nm). Reproduced with permission. © 2012, Elsevier [11]

Thus, the envelope functions' DFT [12] is $(\psi_{\boldsymbol{\kappa}}) = (T_\gamma^+(2\pi\kappa_x d/L, 2\pi\kappa_y d/L, 2\pi\kappa_z d/L))(\phi_{\boldsymbol{\kappa}}) = (T_{\gamma,\boldsymbol{\kappa}}^+)(\phi_{\boldsymbol{\kappa}})$. In this expression, $(\psi_{\boldsymbol{\kappa}})$ is a column vector of four components $\psi_{\nu,\boldsymbol{\kappa}}$ (with ν referring to the associated GBF) and $(T_{\gamma,\boldsymbol{\kappa}}^+)$, of terms $T_{\gamma,\boldsymbol{\kappa}}^{+(\nu,u)}$, is the Hermitical conjugate (and inverse) of the square matrix described in Table 3.1, which is associated to the projection onto the GBFs of states of different bands. In fact, for a given eigenvalue, only one band is involved and a single term of the column vector $(\phi_{\boldsymbol{\kappa}})$ is not zero, so that $\psi_{\nu,\boldsymbol{\kappa}} = T_{\gamma,\boldsymbol{\kappa}}^{(u,\nu)}\phi_{u,\boldsymbol{\kappa}}$ (note that, since all the terms are real, the Hermitic conjugation is performed by permuting the indices). Once $\psi_{\nu,\boldsymbol{\kappa}}$ is known, $\Psi_{\nu.\boldsymbol{n}}$ can be obtained by application of the IDFT. Note that the terms $T_{\gamma,\boldsymbol{\kappa}}^{(u,\nu)}$ are to be calculated at the coordinates $(2\pi\kappa_x d/L, 2\pi\kappa_y d/L, 2\pi\kappa_z d/L)$. The ratio d/L is to be used because (T_γ^+) appears as a function $\left(T_\gamma^+(K_x, K_y, K_z)\right)$ in Table 3.1, and not of (k_x, k_y, k_z).

The calculation lattice element-of-volume l^3 described in the preceding paragraphs is not to be confused with the crystal unit cell volume Ω_{cell}, used, e.g. in discussing the integral factorization rule of Sect. 2.1.1. Actually, the smaller l^3 the more accurate (and time consuming) our calculations are but it has to be bigger than Ω_{cell}. In many of our calculations $l = 1.5$ nm, which is about 3 times the side of the atomic unit cell and l^3 is about 27 times Ω_{cell}. According to Eq. (3.20), the largest value of $k_x = Kx/d$ is $2\pi(N - 1)/2L \cong \pi/l$ (and similarly for the other coordinates) to be compared to the boundary of the 1st Brillouin zone situated at π/a (being a the side of the unit cell). The parabolic character of the bands assumed in the EKPH approximation ceases to be valid if k_x approaches to the edge of the Brillouin zone. As expressed before, in our calculations it remains in the central 1/27 of the Brillouin zone volume (more precisely, 1/27 of the volume of the cube containing the Brillouin Zone).

In summary, in the EKPH, the calculation of the envelope function proceeds along the following scheme [11, 12]: (a) the diagonalized Hamiltonian eigenfunction is calculated; (b) its DFT is calculated; (c) it is multiplied (term by term for each $\boldsymbol{K}$) by the corresponding matrix $(T_{\alpha,\boldsymbol{\kappa}}^+)$ element; (d) the IDFT is calculated. The rows of the $(T_{\gamma,\boldsymbol{\kappa}})$ matrix [the columns of $(T_{\gamma,\boldsymbol{\kappa}}^+)$] correspond to the different energy band eigenvalues (1, 2, 3 and 4 for the *cb*, *lh*, *hh* and *so* respectively). The columns refer to the projections on the *X*-, *Y*-, *Z*- and *S*-GBFs.

Using the integral factorization rule of Eq. (2.1),

$$\langle \Xi(\boldsymbol{r}) \mid \Xi(\boldsymbol{r})\rangle = 1 = \sum_{\nu,\nu'} \langle u_{0,\nu}(\boldsymbol{r}) \mid u_{0,\nu'}(\boldsymbol{r})\rangle\langle \Psi_\nu(\boldsymbol{r}) \mid \Psi_{\nu'}(\boldsymbol{r})\rangle = \sum_{\nu} \langle \Psi_\nu(\boldsymbol{r}) \mid \Psi_\nu(\boldsymbol{r})\rangle \tag{3.27}$$

The later equality makes use of the orthonormal properties of the set $\left|u_{0,\nu}(\boldsymbol{r})\right\rangle$. The integral $\langle \Psi_\nu(\boldsymbol{r}) \mid \Psi_\nu(\boldsymbol{r})\rangle$ can be considered the projection of $|\Xi(\boldsymbol{r})\rangle$ onto the corresponding GBF. In the discretized scheme used, Eq. (3.27) is written as the sum (note that, by definition, $\Psi^*_{v,n}\Psi_{v,n}$ contains an element of volume, $\Delta\Omega$, so that the sums are approximate integrals)

$$\sum_{v}\langle\Psi_v(\boldsymbol{r}) \mid \Psi_v(\boldsymbol{r})\rangle = \sum_{v,n}\Psi^*_{v,n}\Psi_{v,n} = \sum_{v,\boldsymbol{\kappa}}\psi^*_{v,\boldsymbol{\kappa}}\psi_{v,\boldsymbol{\kappa}} = \sum_{v,\boldsymbol{\kappa}}T^{(u,v)}_{\gamma,\boldsymbol{\kappa}}T^{*(u,v)}_{\gamma,\boldsymbol{\kappa}}\phi^*_{u,\boldsymbol{\kappa}}\phi_{u,\boldsymbol{\kappa}}$$
$$= \sum_{\boldsymbol{\kappa}}\phi^*_{u,\boldsymbol{\kappa}}\phi_{u,\boldsymbol{\kappa}} = 1 \tag{3.28}$$

The second equality describes the Parseval's Theorem of the Fourier Transforms. The fourth equality derives from the fact that the rows of $(T_{\gamma,\boldsymbol{\kappa}})$ are normalized for any value of $\boldsymbol{\kappa}$. In the last equality we express that $\boldsymbol{\Phi_n}$ has been normalized. Some problems of normalization at the origin (where the VB is triply degenerate) may appear; they are treated carefully in [11].

The envelope functions corresponding to the diagonalized Hamiltonian solutions in Fig. 3.6 are presented in Fig. 3.7 for a Givens rotation of $(T_{\gamma,\boldsymbol{\kappa}})$ with $\gamma = 0.3$.

In Fig. 3.8, we represent the projection onto the different GBFs of the *cb*(1,1,1) and the *hh*(1,2,1) states versus the Givens rotation parameter γ (the transition involved in the sub-bandgap absorption threshold; *hh*(1,1,1) → *cb*(1,1,1) is forbidden, as explained in Sect. 3.3). Solid lines refer to the transformation matrix studied so far: the row corresponding to the *hh* states is the third in Table 3.1. We call it the standard configuration. We observe that the *cb* state is almost totally projected onto the *S*-GBF (around 96 %); the remaining envelopes are small but still visible. It is mainly a *cb* state and this justifies the commonly used one-band approximation to study it. This statement is thought to be valid for other *cb* states. For low γ, the *hh* state is essentially projected onto the *X*- and *Y*-GBFs and to a lesser extent onto the Z-GBF. The projection onto the *S*-GBF is very small for $\gamma < 0.3$. This means that the *hh* state is almost totally projected onto the *X*-, *Y*- and *Z*-GBFs (which are eigenvalues of the VB). For $\gamma > 0.3$, the projection onto the *S*-GBF is significant, and we may say that the wavefunction contains VB and CB characteristics. If the state *hh*(1,2,1) is changed into *hh*(2,1,2), the projection onto the *X*- and *Y*-envelopes are interchanged, as expected. The rest remains unchanged.

The standard configuration, which associates the third row in Table 3.1 to the *hh* eigenvalue (remember that rows and columns are interchanged in the transposed matrix) and the forth to the *so* eigenvalue, is arbitrary. The opposite option might have been chosen and the fourth row might have been associated to the *hh* eigenfunction. This is called here the interchanged configuration. This is presented in Fig. 3.8 using dashed lines. Again, for low γs, the projection on the *S*-GBF is negligible.

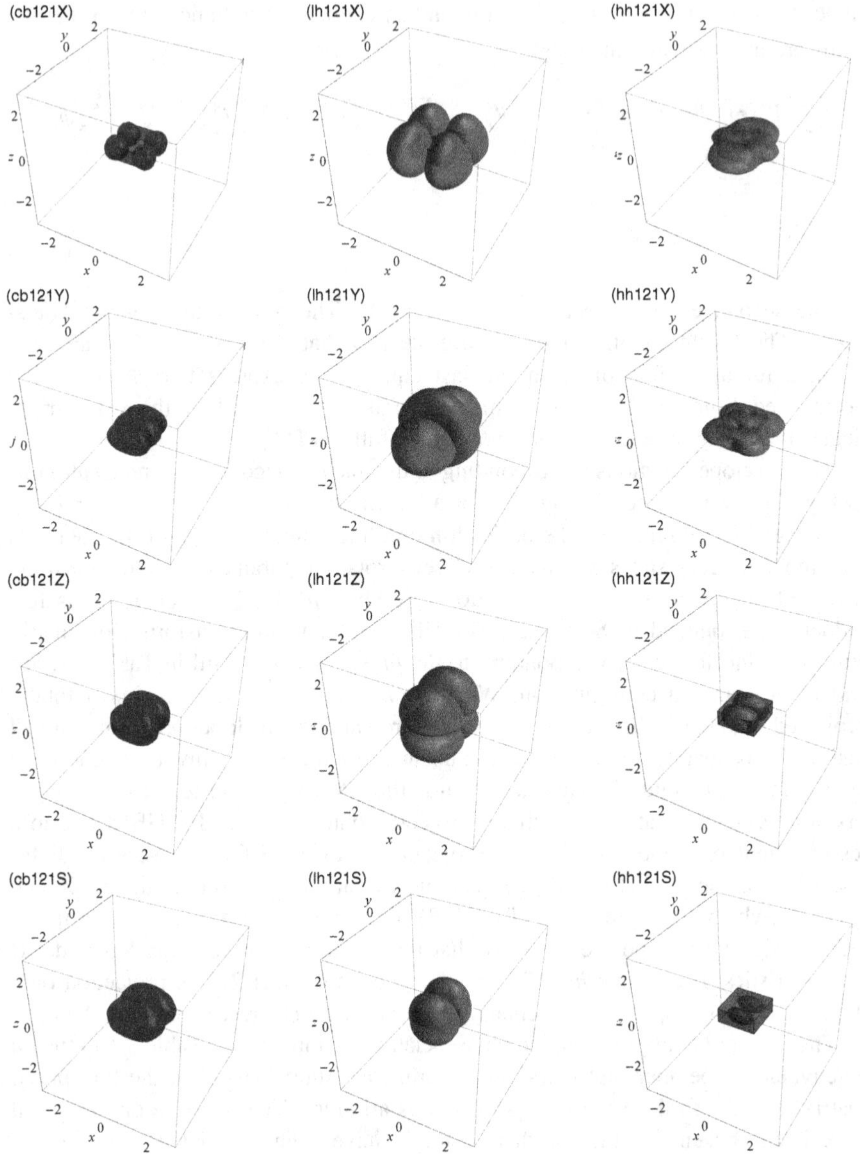

Fig. 3.7 Equal density of probability contours corresponding the same values as in Fig. 3.6 for the states (1,2,1) using a transformation matrix with a Givens rotation $\gamma = 0.3$. *Blue* for CB states and *red* for VB states. The band corresponding to each picture (*cb*, *lh* and *hh*) is indicated in the labels, together with the (*X*-, *Y*-, *Z*- and *S*) envelope it corresponds to. Space coordinates are in units of d (d = 10 nm). Reproduced with permission. © 2012, Elsevier [11]

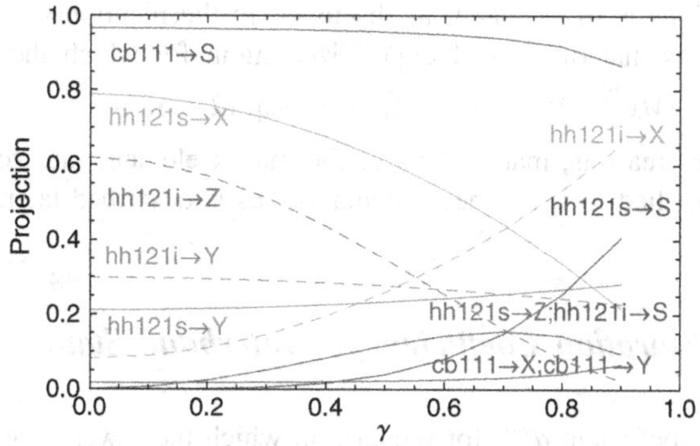

Fig. 3.8 Projection on the GBFs of the *cb*(1,1,1) and *hh*(1,2,1) states. *Solid lines*, standard configuration (a "s" appears in the labels). *Dashed lines*, interchanged configuration (an "i" appears in the labels). Reproduced with permission. © 2012, Elsevier [11]

3.2.3 The Photon-Absorption Elements of Matrix

The dipole matrix ruling the photon-electron interaction is

$$\begin{aligned}\langle\Xi|\epsilon\cdot\boldsymbol{r}|\Xi'\rangle &\cong \sum_{v,v'}\langle u_{0,v}\mid u_{0,v'}\rangle\langle\Psi_v|\epsilon\cdot\boldsymbol{r}|\Psi'_{v'}\rangle = \sum_{v}\langle\Psi_v|\epsilon\cdot\boldsymbol{r}|\Psi'_v\rangle\\ &= \langle\Psi_S|\epsilon\cdot\boldsymbol{r}|\Psi'_S\rangle + \langle\Psi_X|\epsilon\cdot\boldsymbol{r}|\Psi'_X\rangle + \langle\Psi_Y|\epsilon\cdot\boldsymbol{r}|\Psi'_Y\rangle + \langle\Psi_Z|\epsilon\cdot\boldsymbol{r}|\Psi'_Z\rangle\end{aligned} \tag{3.29}$$

where ϵ is the light polarization vector. In this equation, use has been made of the integral factorization rule of Eq. (2.1) as well as the orthonormal properties of the GBFs. Thus, under this approximation, only the envelope functions play a role and we can ignore the GBFs, which are always difficult to know. Exact expression of the absorption coefficients in function of the matrix elements may be found in [11, 12, 21, 22]. We can define the polarization by the Euler angles(φ, θ) characterizing its direction, as done in Sect. 2.4.1. Thus

$$\langle\Xi|\varepsilon\cdot\boldsymbol{r}|\Xi'\rangle = \cos\varphi\sin\theta\langle\Xi|x|\Xi'\rangle + \sin\varphi\sin\theta\langle\Xi|y|\Xi'\rangle + \cos\theta\langle\Xi|z|\Xi'\rangle \tag{3.30}$$

and the same for the envelopes composing the element of matrix. We use the following nomenclature,

$$\begin{aligned}Mx &\equiv \langle\Xi|x|\Xi'\rangle = \langle\Psi_X|x|\Psi'_X\rangle + \langle\Psi_Y|x|\Psi'_Y\rangle + \langle\Psi_Z|x|\Psi'_Z\rangle + \langle\Psi_S|x|\Psi'_S\rangle\\ &\equiv MxX + MxY + MxZ + MxS\end{aligned} \tag{3.31}$$

and similarly for the y and z polarization components.

The absorption coefficient is proportional to the squared module of the matrix element. Furthermore, in many cases, the light is unpolarized, and the average of a set of polarizations is to be taken, as discussed in the mentioned subsection. An interesting case is that of vertical unpolarized light, for which the absorption is proportional to $\left(|Mx|^2+|My|^2\right)/2$ [22], as in Eq. (2.75).

For symmetry reasons, many terms of the matrix elements are zero and group theory can be applied to spare their calculation, as is described later, in Sect. 3.3.

3.2.4 The Absorption Coefficient for Interband States

The absorption coefficient α^{max} for the case in which the lower state is totally full and the upper state empty is given by Eq. (2.70), which can be modified later by the occupancy factor of the states. The α^{max} are the quantities that will be the starting point for a realistic detailed balance study [21].

In this subsection, we follow reference [23] closely.

Figure 3.9a (left panel) shows a simplified spatial band diagram for a $16 \times 16 \times 6\ \text{nm}^3$ QD with the other relevant parameters in the SOTA prototype. A Givens rotation of $\gamma = 0.3$ is used. Horizontal lines represent the confined state energy of the *cb*, heavy holes (*hh*), light holes (*lh*) and split-off band (*so*) states. Within the CB potential well, there are three distinct bound state energy levels, which are separated from the rest of the *cb* states. Within this study, these are referred to as IB states. The bound states (BS) whose energy is above the matrix CB band edge [so-called virtual bound states (VBSs)] are here considered part of the device CB. Due to their high effective mass, the VB potential pedestal contains a near-continuum of *hh* states. Because of the close packing of these levels, they are considered not as a separate band but as an extension of the device VB into the forbidden band. Note that some CB and VB virtual bound states extend deep into the respective bands; they are not shown in this or in later figures. Note also that the level denoted as IB(2,1,1) (alternatively notation: $|\text{IB}211\rangle$) actually refers to the degenerate pair of states IB (2,1,1)/IB(1,2,1).

We are interested in the net contribution of transitions from all VB (*hh*, *lh* and *so*) states to each individual IB state. An absorption coefficient has been calculated for each final IB state by

$$\alpha_{VB\to j}^{max} = \sum_{i\,\epsilon\,all\ VB\ states} \alpha_{i\to j}^{max} \tag{3.32}$$

where *i* denotes the initial and *j* the final state. The absorption coefficients are plotted in Fig. 3.9a (right panel) for all final IB states. Although in each case the absorption coefficient sums contributions from the 169 VB states, distinct peaks can be seen pertaining to a few dominant transitions. Taking $\alpha_{VB\to \text{IB}(1,1,1)}^{max}$ as an example, the dominant transitions are those from the *hh*(2,1,1), *hh*(4,1,1), *hh*(6,1,1) and *lh*(2,1,1) states (the latter three have degenerate counterparts *hh*(1,4,1), *hh*(1,6,1) and *lh*(1,2,1), which make an equal contribution). Each peak in the absorption curve is labeled with

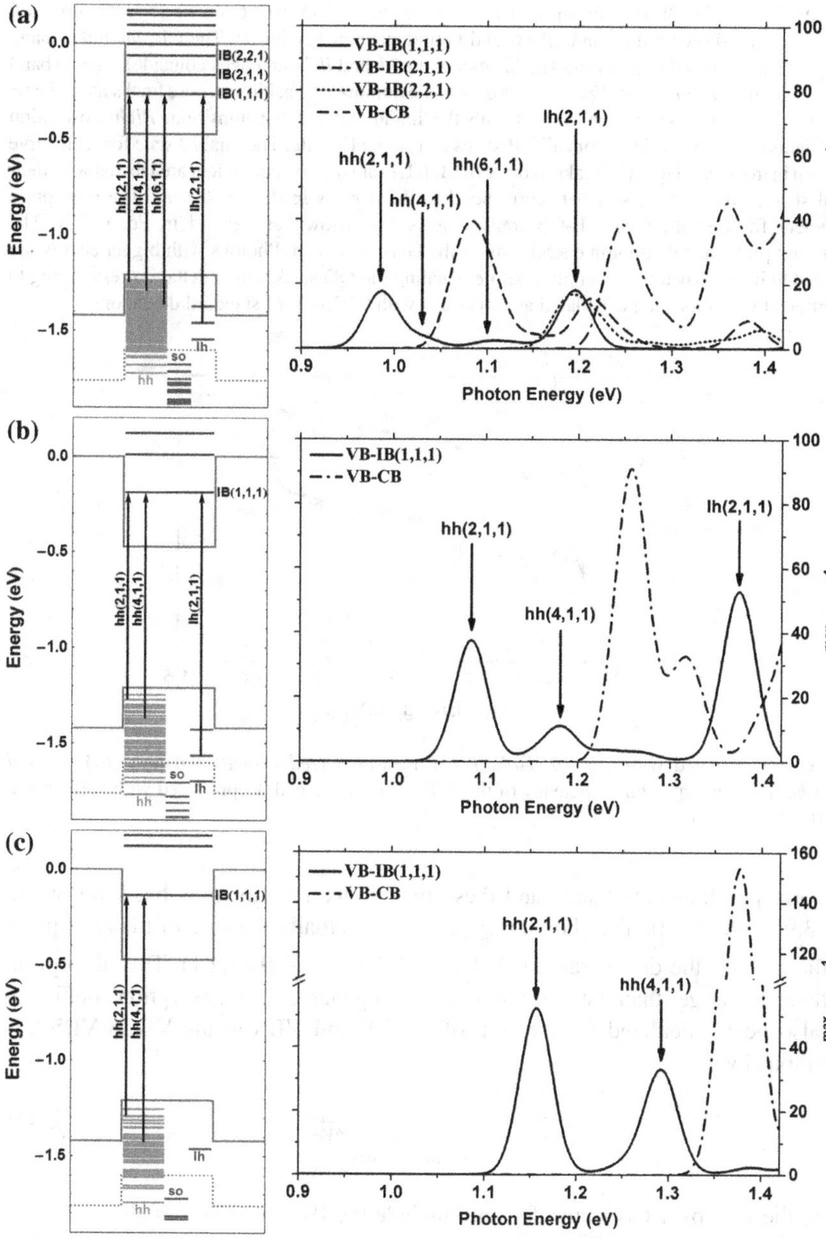

(a)
Energy (eV)
IB(2,2,1)
IB(2,1,1)
IB(1,1,1)
hh(2,1,1)
hh(4,1,1)
hh(6,1,1)
lh(2,1,1)
so
hh
lh
VB-IB(1,1,1)
VB-IB(2,1,1)
VB-IB(2,2,1)
VB-CB
lh(2,1,1)
hh(2,1,1)
hh(6,1,1)
hh(4,1,1)
Photon Energy (eV)
α^{max} (cm^{-1})
(b)
Energy (eV)
IB(1,1,1)
hh(2,1,1)
hh(4,1,1)
lh(2,1,1)
so
lh
hh
VB-IB(1,1,1)
VB-CB
hh(2,1,1)
lh(2,1,1)
hh(4,1,1)
Photon Energy (eV)
α^{max} (cm^{-1})
(c)
Energy (eV)
IB(1,1,1)
hh(2,1,1)
hh(4,1,1)
lh
so
hh
VB-IB(1,1,1)
VB-CB
hh(2,1,1)
hh(4,1,1)
Photon Energy (eV)
α^{max} (cm^{-1})

◀**Fig. 3.9** Results for QD dimensions of **a** 16 × 16 × 6 nm^3, **b** 10 × 10 × 6 nm^3, **c** 8 × 8 × 6 nm^3. *Left* Band diagrams showing the band offsets and bound state energy levels. The CB, *hh* and *lh* band edges are shown as *solid grey lines* (in this model the *hh* and *lh* band edges coincide). The *so* band edge is shown as a *dotted grey line*. Arrows denote the dominant transitions whose final state is the IB (1,1,1) state. These *arrows* are labeled with the initial state of the transition. *Right* absorption coefficients for the transitions from all VB states to a single IB state. The final IB state for each curve is shown in the figure legends. Peaks in the VB-IB(1,1,1) absorption coefficient are labeled with their initial state in the VB; these labels correspond to the *arrows in the left figures*. The absorption coefficient for bound-bound VB-CB transitions is also shown as defined in Eq. (3.33). The absorption plots include photon energies up to the GaAs bandgap. Photons with bigger energy are assumed to be absorbed by the emitter before reaching the QD stack. Dirac deltas corresponding to the different transitions are calculated as Gaussians with 0.025 eV of standard deviation

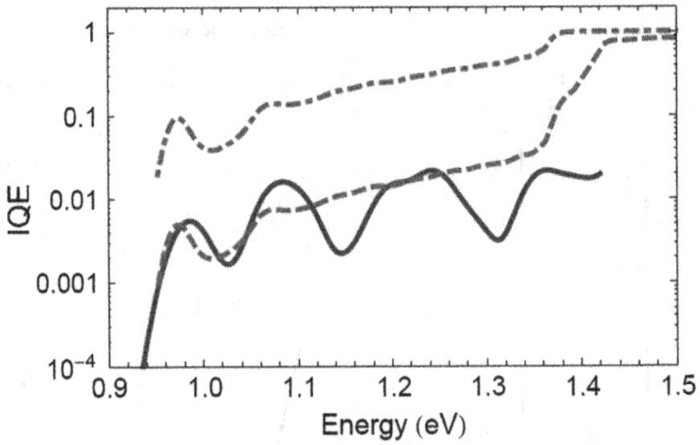

Fig. 3.10 Calculated with γ = 0.2, (*solid/blue*), measured (*red/dashed*) and projected (*sienna/dotdashed*) internal quantum efficiency of the SOTA prototype cell. Reproduced with permission. © 2012, Elsevier [11]

the corresponding initial state and these transitions are shown as black arrows in Fig. 3.9a (left panel). The shown $\alpha^{max}_{VB\to IB(2,1,1)}$ is actually the sum of the absorption coefficients for the degenerate final states IB(2,1,1) and IB(1,2,1). This absorption coefficient is larger than the others due to this degeneracy. An absorption coefficient has also been calculated for the sum of the BS and VBS in the VB to VBS CB transitions by

$$\alpha^{max}_{VB\to CB} = \sum_{j\,\epsilon\text{ all CB states}} \alpha^{max}_{VB\to j} \tag{3.33}$$

where the sum over CB states does not include the IB states. This is also shown in the figure.

The calculations have been repeated for a 10 × 10 × 6 nm^3 QD and for an 8 × 8 × 6 nm^3 QD. The results are shown in Fig. 3.9b and c. Note that Fig. 3.9c has a break in the *y* scale to allow all graphs to be plotted with the same scaling. For these dimensions, there is only a single IB energy level. The possible benefits of such a configuration are discussed in [24]. Regarding the VB-IB transitions, some further observations can be made. Firstly, there is a general increase in the absorption coefficient on decreasing the QD width. Secondly, the absorption peaks move to higher energy due to the bound states moving deeper into their respective bands. This is most pronounced for the transition whose initial state is *lh*(2,1,1). Finally, for the $10 \times 10 \times 6$ nm^3 QD, the *hh*(6,1,1) absorption peak is no longer present, and, for the $8 \times 8 \times 6$ nm^3 QD, the *lh*(2,1,1) absorption peak is no longer present. This is due to the disappearance of the respective states in the VB at these QD widths. This will have some negative effect on the overall VB-IB photocurrent.

3.2.5 Experimental Check

In the SOTA prototype, the sub-bandgap absorption is obtained by adding all the curves in Fig. 3.9a after multiplying by 0.8 the curve corresponding to the VB → *cb*(1,1,1) absorption to account for the approximate 20 % filling by doping of this level.

Then, the IQE can be determined under the assumption that all the electron-hole pairs generated are extracted to produce photocurrent. The IQE efficiency is then give by

$$IQE = 1 - \exp(-\alpha W_{IB}) \tag{3.34}$$

where W_{IB} is the thickness of the IB region, in this case of 2.4 μm.

This is presented in Fig. 3.10 together with the IQE measured [12] in the SOTA prototype. The agreement with experiment is reasonably high. The energy at which the real transitions happen are coincident and the theoretical and experimental values of the IQE are similar. For extraterrestrial radiation, the calculated sub-bandgap current (truncated at 1.38 eV, before the wetting layer absorption, actually a quantum well) is 0.208 mA cm^{-2} and the measured one is 0.266 mA cm^{-2} (reflection excluded in both cases). This similarity happens also for two more cells of different QD sizes we have tested and this is considered a validation of the model. It has to be stressed that nothing has been fitted to obtain this result. The QD size is measured by TEM and the density of QDs by AFM [25]. The potential offsets for the CB and the VB are those calculated for the InAs/GaAs system by ab initio calculations with 2×10^6 atoms supercell using QDs of not very different (1/2) size [14]. The InAs effective masses are taken from the literature [19]. However, while the effective mass of the InAs (QD material) is used for the calculation of the wavefunctions, the diagonalization matrix is calculated using the effective mass of the GaAs (the barrier material) and the arithmetic mean of the QD and barrier bandgaps.

The calculated IQE increases with energy, but not as much as the experimental one. Leaving aside the effect of the wetting layer, not studied in this book, we have neglected the transitions form the bound states in the QD to the extended states in the CB and VB that we consider not very important (this is confirmed later in this work) and the effect of the reduction of the energy of the VBS due to the exact calculation of the box shaped potential studied in Sect. 2.3. This would bring probably more transitions to the sub-bandgap high energy region. These transitions are now beyond the bandgap and are not considered. Our calculations are wavier than the measurements. This may be due the approximation of the Dirac deltas by a Gaussian with 0.025 eV of standard deviation. This is assumed to reflect the QD variation of size and this may be correct for the lower levels. It is however too small for larger levels, which shift more for the same change in size. In this way, the Gaussian may be smeared without changing its integral value but showing a less wavy aspect. More transitions will come into the account if the second order approximation in the factorization of integrals rule is taken into account. But this requires information about GBFs, which is not easily available.

Some practically important conclusions can also be extracted. The sub-bandgap absorption in a QD IB solar cell is low because the QDs absorb poorly: the IB (*cb*) states are basically *S* functions, and the *hh* states lack this component (because $T_{0,\boldsymbol{\kappa}}^{(3,4)} = 0$) when the Givens rotation is 0 or have it very weak for other reasonable Givens rotations. Nevertheless, reasonably good sub-bandgap absorption would be produced if we were to multiply the number of QDs by 20. This can be achieved by increasing the coverage factor to $F_s = 0.5$ [see Eq. (2.70)] and reducing the spacers between QDs to 20 nm while preserving the present IB region thickness (2.4 mm). All this seems feasible [26, 27]. The tunneling that has been avoided with the use of 80 nm spacers should not be an issue in a neutral region and the spacers may be left thick in the space charge zone (or field damping layers may be added [28]). By integration of the IQE, this would give a reasonable sub-bandgap current density of 4.433 instead of 0.208 mA cm^{-2} to add to the value from the GaAs bandgap, of about 28.9 mA cm^{-2}. Further sub-bandgap current density can be generated if light trapping structures are added to the cell [29]. The subject will be treated more in detail in Chap. 4.

As presented in Fig. 3.8, the projection of the eigenfunction in the different GBFs depends on the Givens rotation γ. In [11], the calculations (which are somewhat different to those in [12] because they take into consideration the effect of the symmetry) have been reproduced for different Givens rotations. The results are in Fig. 3.11. It can be seen that the best integrated fitting is with $\gamma = 0.2$ in the standard configuration and with $\gamma = 0.1$ in the interchanged configuration. To the eye, looking at the whole IQE plot, the best fitting is the former one. The differences with [12], on which the discussion above is based, are also very small.

Along the entire chapter, unless otherwise specifically said, the data of the table in Fig. 3.4 are used in the calculation of the diagonalized wavefunctions. The effective masses correspond to the QD material (InAs). However, the data in Table 3.2 are used in the calculation of the transformation matrix.

The effective mass and the bandgap of the host material (GaAs) are used in the calculation of parameter *B*, appearing in the Hamiltonian of Fig. 3.1. This parameter is inversely proportional to the value of the normalizing size *d*. In the calculation of the (*T*) matrix, average of the effective masses in the host (GaAs) and in the QD (InAs) materials is used, in addition of B, which is calculated as already said.

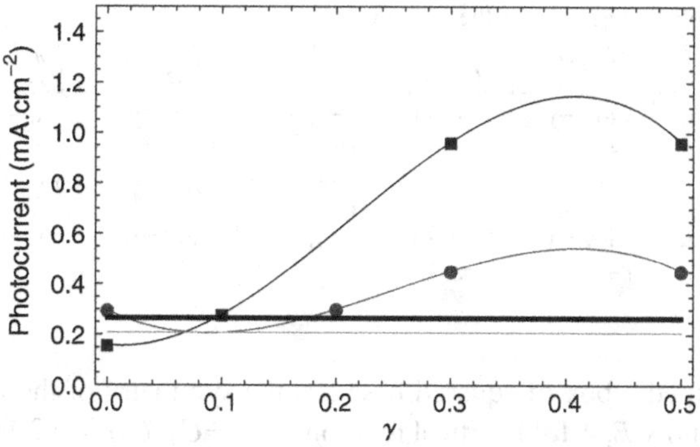

Fig. 3.11 Sub-bandgap photocurrent calculated for several values of γ for the standard configuration (*blue dots*) and for the interchanged configuration (*purple squares*). *Fitted lines* are mainly eye guides. The *thick black horizontal line* represents the experimental sub bandgap photo-current in reference [11] and the *orange line* the photo-current calculated in reference [12]. Reproduced with permission. © 2012, Elsevier [11]

Table 3.2 Data used for the calculation of the (*T*) matrix

m_{cb} (in B)/m_0	E_g (in B)/eV	B (d = 1 nm)/eV	E_g in (T)/eV
0.4535	1.42	1.06718	1.0785

The EKPH model has an inherent ambiguity concerning the choice of parameters for the (*T*) matrix. There is not a convincing theoretical argument for this choice, beyond the satisfactory fitting with experimental quantum efficiency. We want to stress that the (*T*) matrix is not used in the spectrum determination, only in the calculation of the envelope functions, and therefore of the optical behavior. A convincing determination of the (*T*) elements can be found in Chap. 7 when the four-band Luttinger Kohn Hamiltonian is described.

3.3 Application of the Group Theory to the Envolvent Functions and Matrix Elements

Reference [11] is followed here closely.

3.3.1 The QD Symmetry

InAs QDs in GaAs grown in the Stransky-Krastanov growth mode form strongly truncated quadrangular pyramids. They induce a change of the material CB and VB. In our simplifying effort, we assume the QD to be a squat parallelepiped of squared base and adopt a mesoscopic potential box of parallelepipedic shape.

Table 3.3 Symmetry operations of the D_{4h} group

E	$I \cdot C''_{x2}$	$I \cdot C''_{y2}$	$I \cdot C_2$	I	C''_{x2}	C''_{y2}	C_2
$+x$, $+y$, $+z$	$-x$, $+y$, $+z$	$+x$, $-y$, $+z$	$+x$, $+y$, $-z$	$-x$, $-y$, $-z$	$+x$, $-y$, $-z$	$-x$, $+y$, $-z$	$-x$, $-y$, $+z$
$I \cdot C'_{ds2}$	$-C_4$	C_4	C'_{dp2}	C'_{ds2}	$I\cdot(-C_4)$	$I\cdot C_4$	$I \cdot C'_{dp2}$
$+y$, $+x$, $+z$	$-y$, $+x$, $+z$	$+y$, $-x$, $+z$	$+y$, $+x$, $-z$	$-y$, $-x$, $-z$	$+y$, $-x$, $-z$	$-y$, $+x$, $-z$	$-y$, $-x$, $+z$

A parallelepipedic box of squared base has the symmetry of the D_{4h} group that includes the identity E, 4-fold vertical rotations (C_4, $-C_4$, C_2) two 2-fold rotations C' around the base diagonals, two 2-fold rotations C'' around the x and y axes and the product of the preceding transformations times the inversion about the center I. In total, this group has 16 symmetry operations. Table 3.3 shows all the cited operations and how the position vector (x, y, z) is transformed by them.

Any function, no matter how complicated it might be, can generate a functional vector space in the following way: the function is modified by the symmetry operations and the new functions generated are the basis of such a space, whose dimension is, at most, the number N_G of operations of the group (16 in this case). However, for certain functions, some of the transformed functions are the original function or linear combinations of some few transformed functions, so leading to spaces of lower dimension.

Each symmetry operation G of the group $\mathbf{G}$ induces a transformation in the functional vector space that is represented by a matrix Γ_G. These matrices constitute a representation of the group $\mathbf{G}$. If all the transformed functions are linearly independent, the dimension of the matrices is N_G and the representation is called the regular representation. If the function generating the representation is highly symmetric, the dimension of the functional space generated may be very small, even one. In the latter case, the function is invariant under the group operations with all matrices of dimension one with a single matrix element of value 1 (the identical representation) or at most (if the function is real) ± 1.

Sometimes, linear combinations of the initial basis functions lead to matrices that are diagonal by blocks, each block corresponding to smaller dimension representations. Actually, in finite groups, all the representations are unitary [30, vol. 2, p. 945]. This means that the matrices are unitary (the lines and columns are orthonormal vectors), and a basis, formed of linear combinations of the initial basis, can be found in which the representation matrices are diagonal in blocks. Such matrices are called the direct sum of the diagonal blocks. Additionally, a basis can be found where these diagonal blocks cannot be further reduced into smaller blocks: they are called irreducible representations [30–33]. We say that the representation $\boldsymbol{\Gamma}$ has been developed into the direct sum of irreducible representations $\boldsymbol{\Gamma}^{(i)}$; that is, $\Gamma_G = \sum_i \Gamma_G^{(i)}$.

Table 3.4 Table of characters of the D_{4h} group of symmetry

Ch. D_{4h}	E	C_2	$2C_4$	$2C'_2$	$2C''_2$	I	IC_2	$2IC_4$	$2IC'_2$	$2IC''_2$	
A_{1g}	1	1	1	1	1	1	1	1	1	1	x^2+y^2, z^2
A_{1u}	1	1	1	1	1	−1	−1	−1	−1	−1	
A_{2g}	1	1	1	−1	−1	1	1	1	−1	−1	
A_{2u}	1	1	1	−1	−1	−1	−1	−1	1	1	z
B_{1g}	1	1	−1	1	−1	1	1	−1	1	−1	x^2-y^2
B_{1u}	1	1	−1	1	−1	−1	−1	1	−1	1	
B_{2g}	1	1	−1	−1	1	1	1	−1	−1	1	xy
B_{2u}	1	1	−1	−1	1	−1	−1	1	1	−1	
E_g	2	−2	0	0	0	2	−2	0	0	0	(xz, yz)
E_u	2	−2	0	0	0	−2	2	0	0	0	(x, y)

The traces χ (sum of the diagonal elements) of the representation matrices are called the characters of the representation. The table of characters of the irreducible representations $\chi^{(i)}$ of the D_{4h}-symmetry group, as taken from the literature [31, 32], is presented in Table 3.4. In the table, the irreducible representations and the symmetry operations grouped in classes (whose exact meaning is not relevant here) appear. In the rightmost column, some basis functions generating irreducible representations that are frequently found in practical applications are listed according to their irreducible representation.

Note that the lines are orthogonal vectors of dimension 16 if the classes are split into the symmetry operations they are formed of (e.g. $2C''_2$ is split into $C''_{x,2}$ and $C''_{y,2}$, both with the same character that depends on the representation). The square of the norm of these vectors is N_G (=16). This is the test to know if a representation is irreducible (if not, the square of the norm is a multiple of N_G).

Once the characters χ of a given representation are known, its development as the direct sum of irreducible representations can involve the repetition of $h^{(i)}$ times each irreducible representation. The formula for the $h^{(i)}$ coefficients is [31, p. 454]

$$h^{(i)} = \sum_{G=1}^{N_G} \chi_G \chi_G^{(i)*} \Big/ N_G \tag{3.35}$$

where the index (i) refers to the irreducible representation and the index G corresponds to the group operation, (note: the classes must be split into their symmetry operations by repeating the same character as many times as the number of elements of the class, in the first line of Table 3.4). The asterisk represents complex conjugation (unnecessary for D_{4h} because the characters are real).

3.3.2 *Theorems of Group Theory*

For each symmetry operation, the characteristic of the Cartesian product of several representation matrices is the ordinary product of their characteristics. Additionally, the characteristic of the direct sum of several representation matrices is the ordinary sum of their characteristics.

The necessary condition for a quantum-mechanical-operator element of matrix to be non-zero is that the Cartesian product representation of the operator and of one of the two wavefunctions and the representation of the remaining wavefunction contain at least one irreducible representation in common. This is equivalent to saying that the Cartesian product of the two wavefunction representations and that of the operator contain the identical representation at least once [31]; the matrix element will be zero if this is not the case. This can be easily verified by calculating the characters of the triple Cartesian product representation and by using Eq. (3.35) to calculate the number of times the identical representation is included. If the signed sum of these characters is zero (it is included zero times), the transition is forbidden. Otherwise, the result must be a multiple of N_G (this can be used as a check) and the transition is not forbidden by symmetry considerations; but the matrix element can still be zero.

3.3.3 *Relating the r and k Point-Symmetries*

Repeating what has already been said, in the k·p method, the envelope functions result from a plane wave development as indicated in Eq. (3.12). In practice, the assumption of a finite number of plane waves results in the determination of the wavefunction at a finite number of nodes of a calculation cubic lattice. The rest of the points, if needed, are to be interpolated. Mathematically speaking, the coefficients $\psi_v(2\pi\boldsymbol{\kappa}/L)$ of the development in Eq. (3.12), that we shall also call $\psi_{v,\boldsymbol{\kappa}}$, are the Discrete Fourier Transform (DFT) of the values of $\Delta\Omega^{1/2}\,\Psi_v(\boldsymbol{r})$ calculated at the lattice nodes where $\Delta\Omega = (L/N)^3 = l^3$ is the element of volume associated to a calculation node and, obviously,[1] $\Omega = \Delta\Omega \times N^3 = L^3$. The values of $\Delta\Omega^{1/2}\Psi_v(\boldsymbol{n}L/N)$ at these nodes ($\boldsymbol{n}$ like $\boldsymbol{\kappa}$ are triplets of integers) are called $\Psi_{v,\boldsymbol{n}}$ and they are the inverse transform (IDFT) of $\psi_{v,\boldsymbol{\kappa}}$. Sometimes, the band sub-index v is dropped for simplicity. Therefore, with the simplified notation, the DFT and the IDFT are defined respectively as

[1] Ω is here the volume of the region considered for calculations; it is the region where the wavefunctions are expected to be non-negligible. The same symbol is used for an arbitrary large volume used in quantum mechanics where all the wavefunctions lay. It is also used to discretize the continuum spectrum. They may or may not be the same.

$$\psi_{\kappa} = \frac{1}{N^{3/2}} \sum_{n_x=-(N-1)/2, n_y=-(N-1)/2, n_z=-(N-1)/2}^{(N-1)/2,(N-1)/2,(N-1)/2} \Psi_{n} e^{-2\pi i \kappa \cdot n/N}$$
$$\Psi_{\mathrm{n}} = \frac{1}{N^{3/2}} \sum_{k_x=-(N-1)/2, k_y=-(N-1)/2, k_z=-(Nx-1)/2}^{(N-1)/2,(N-1)/2,(N-1)/2} \psi_{\kappa} e^{2\pi i \kappa \cdot n/N} \tag{3.36}$$

The sum terms depend on three integer indices as explained above. This is perhaps not the most common way of defining the DFT and the IDFT. It is more frequent to use natural numbers for the indices starting with the term of zero-exponent. Here the zero-exponent is to be found in the middle of the series for each index. The usual definition can be changed into the one used here if, before the transformation, (a) the array of numbers suffers a circular permutation of (N − 1)/2 positions to the left with respect to each index, (b) the ordinary DFT is obtained and (c) the resulting array of numbers is again rotated (N − 1)/2 times to the right for each index.

Let G be a symmetry operation belonging to the group $\boldsymbol{G}$, which is a subgroup of the cube symmetry O_h (as the D_{4h} symmetry is). By definition,

$$G\Psi_{n} \equiv \Psi_{G^{-1}n} \tag{3.37}$$

It can be proven that [33, p. 107]

$$G\psi_{\kappa} = \psi_{G^{-1}\kappa} \tag{3.38}$$

What we want to prove now is that the DFT of any function Ψ_{n}, ψ_{κ}, is transformed by the group G symmetry operations according to the same representation as Ψ_{n}.

Let us now assume that Ψ_{n} is transformed by the symmetry operations of $\boldsymbol{G}$ according to the representation matrices Γ_G (it could be one of the irreducible representations in Table 3.4). Let $\{\Psi_{n}^{i}\}$ be the set of basis functions of this representation. Then,

$$G\Psi_{n}^{p} = \Psi_{G^{-1}n}^{p} = \sum_{q} \Gamma_{G}^{p,q} \Psi_{n}^{q} \tag{3.39}$$

The DFT of $G\Psi_{n}^{p}$ is, by definition

$$G\psi_{\kappa}^{p} \equiv \frac{1}{N^{3/2}} \sum_{n} G\Psi_{n}^{p} e^{-2\pi i \kappa \cdot n/N}$$
$$= \frac{1}{N^{3/2}} \sum_{n} \sum_{q} \Gamma_{G}^{p,q} \Psi_{n}^{j} e^{-2\pi i \kappa \cdot n/N} = \sum_{q} \Gamma_{G}^{p,q} \psi_{\kappa}^{q} = \psi_{G^{-1}\kappa} \tag{3.40}$$

hence the DFT and the IFDT are transformed by the symmetry operations according to the same representation and therefore they have the same table of characters.

3.3.4 Symmetry Considerations in the Calculation of the Dipole Optical Elements

3.3.4.1 Symmetry of the Diagonalized Hamiltonian Eigenfunctions

Let us consider first the diagonalized Hamiltonian eigenfunctions. We look initially at function[2] $|1,1,1\rangle$. It is the product of three even one-dimensional functions, one for each coordinate. Furthermore, the functions for x and for y are exactly the same. Therefore, applying the D_{4h} symmetries, the three-dimensional eigenfunction is always invariant (see in Table 3.3 how the variables are transformed by the different symmetry operations). They follow the identical representation A_{1g}. The same will happen for any eigenfunction whose QNs are of the type $|\mathrm{odd1},\mathrm{odd1},\mathrm{odd}\rangle$, the first two corresponding to the same one-dimensional function, in this case, of even symmetry. The z-eigenfunction is always different because in this direction the potential well has a width $2c \neq 2a = 2b$.

With the same arguments, looking again at Table 3.3, we can see how the eigenfunctions with QNs in the second, third and fourth lines of Table 3.5 are transformed. The symmetry operations leave the eigenfunctions unchanged (character +1) or change their sign (character −1) and they all follow different one-dimensional ($\chi_E = 1$) irreducible representations of the symmetry group.

For the case of $|1,2,1\rangle$, the level is doubly degenerate and the basis functions are $|1,2,1\rangle$ and $|2,1,1\rangle$. The representation is two-dimensional ($\chi_E = 2$), the eigenfunction $|1,2,1\rangle$ may correspond to the representation column vector $\{1,0\}$ and the eigenfunction $|1,2,1\rangle$ to the column vector $\{0,1\}$ (both written as rows for notation simplicity). For, e.g. the symmetry operation IC_4, which transforms the point (x,y,z) into $(-y,x,-z)$, the eigenfunction $|1,2,1\rangle$ is transformed into $|2,1,1\rangle$ and $|2,1,1\rangle$ into $-|1,2,1\rangle$, in the representation space, to the representation vector $\{1,0\}$ corresponds $\{0,1\}$ and to the vector $\{0,1\}$ corresponds the vector $\{-1,0\}$. The matrix of this transformation appears in Table 3.6 under the IC_4 symmetry operation. We can repeat this procedure for the rest of the symmetry operations. The transformation matrices are also in Table 3.6. The characters (which are the traces of the matrices) correspond to the two-dimensional irreducible representation E_u, as presented in Table 3.5. The preceding arguments are also applicable to the eigenfunction of the following line in Table 3.5 whose representation is E_g.

---

[2] Along this book we usually denote eigenstates with the quantum numbers of their separable wavefunctions, sometimes preceded of the band notation; however, in this subsection we use the Dirac notation for state functions for easier differentiation of the representation column vectors associated to these eigenstate functions, which are denoted as a row of numbers in braces.

Table 3.5 Representation associated to the diagonalized Hamiltonian eigenfunctions

One-dimensional quantum numbers (parity of the function opposite to parity of QN)	Column vectors for 1D representations	Irreducible representation
$\|odd1, odd1, odd\rangle$		A_{1g}
$\|odd1, odd1, even\rangle$		A_{2u}
$\|even1, even1, odd\rangle$		B_{1g}
$\|even1, even1, even\rangle$		B_{2u}
$\|odd1, even2, odd\rangle$, $\|even1, odd2, odd\rangle$		E_u
$\|odd1, even2, even)$, $\|even1, odd2, even\rangle$		E_g
$\|odd1, odd2, odd\rangle$	{1, 1},{1, −1}	$A_{1g} + B_{2g}$
$\|odd1, odd2, even\rangle$	{1, 1},{1, −1}	$A_{2u} + B_{1u}$
$\|even1, even2, odd\rangle$	{1, −1},{1, 1}	$A_{2g} + B_{1g}$
$\|even1, even2, even\rangle$	{1, −1},{1, 1}	$A_{1u} + B_{2u}$

Table 3.6 Matrices of the two-dimensional representation subtended by the eigenfunctions $|odd1, even2, odd\rangle$ and $|even2, odd1, odd\rangle$

E	$I \cdot C''_{x2}$	$I \cdot C''_{y2}$	$I \cdot C_2$	I	C''_{x2}	C''_{y2}	C_2
+x, +y, +z	−x, +y, +z	+x, −y, +z	+x, +y, −z	−x, −y, −z	+x, −y, −z	−x, +y, −z	−x, −y, +z
$\begin{pmatrix} 1 & 0 \\ 0 & 1 \end{pmatrix}$	$\begin{pmatrix} 1 & 0 \\ 0 & -1 \end{pmatrix}$	$\begin{pmatrix} -1 & 0 \\ 0 & 1 \end{pmatrix}$	$\begin{pmatrix} 1 & 0 \\ 0 & 1 \end{pmatrix}$	$\begin{pmatrix} -1 & 0 \\ 0 & -1 \end{pmatrix}$	$\begin{pmatrix} -1 & 0 \\ 0 & 1 \end{pmatrix}$	$\begin{pmatrix} 1 & 0 \\ 0 & -1 \end{pmatrix}$	$\begin{pmatrix} -1 & 0 \\ 0 & -1 \end{pmatrix}$
$I \cdot C'_{ds2}$	$-C_4$	C_4	C'_{dp2}	C'_{ds2}	$I \cdot (-C_4)$	$I \cdot C_4$	$I \cdot C'_{dp2}$
+y, +x, +z	−y, +x, +z	+y,−x, +z	+y, +x,−z	−y, −x, −z	+y, −x, −z	−y, + x, −z	−y, −x, +z
$\begin{pmatrix} 0 & 1 \\ 1 & 0 \end{pmatrix}$	$\begin{pmatrix} 0 & -1 \\ 1 & 0 \end{pmatrix}$	$\begin{pmatrix} 0 & 1 \\ -1 & 0 \end{pmatrix}$	$\begin{pmatrix} 0 & 1 \\ 1 & 0 \end{pmatrix}$	$\begin{pmatrix} 0 & -1 \\ -1 & 0 \end{pmatrix}$	$\begin{pmatrix} 0 & 1 \\ -1 & 0 \end{pmatrix}$	$\begin{pmatrix} 0 & -1 \\ 1 & 0 \end{pmatrix}$	$\begin{pmatrix} 0 & -1 \\ -1 & 0 \end{pmatrix}$

The procedure above can also be applied to the $|odd1, odd2, odd\rangle$ eigenfunctions. They also generate two-dimensional representations and the corresponding matrices may be deduced along with the table of characters of the representation. However, they are not irreducible (the sum of the squared of the characters is not $N_G = 16$, but 32). By application of Eq. (3.35), they are found to be the direct sum of two one-dimensional irreducible representations. This appears in the corresponding Table 3.5, line 7. In the column labeled "Column vector for 1D representations", the vectors leading to these one-dimensional representations are registered. In the case we are examining, the $\{1, 1\}$ and $\{1, -1\}$ vectors correspond to the eigenfunction $|odd1, odd2, odd\rangle + |odd2, odd1, odd\rangle$ and $|odd1, odd2, odd\rangle - |odd1, odd2, odd\rangle$. It can be seen that these eigenfunctions transform into themselves, with or without a change of sign, so reflecting the fact that they span a one-dimensional representation. The same happens for all the symmetry combinations in the Table 3.5, lines 8, 9 and 10.

Table 3.7 Irreducible representations followed by the elements of the $(T_{\gamma,A})$ and $(T_{\gamma,B})$ matrices

$\Gamma(T_{\gamma,A}) = \begin{pmatrix} E_u & E_u & A_{2u} & A_{1g} \\ E_u & E_u & A_{2u} & A_{1g} \\ E_u & E_u & 0 & 0 \\ E_g & E_g & A_{1g} & 0 \end{pmatrix}$	$\Gamma(T_{\gamma,B}) = \begin{pmatrix} 0 & 0 & A_{1g} & A_{2u} \\ 0 & 0 & A_{1g} & A_{2u} \\ E_g & E_g & A_{1g} & A_{1g} \\ E_u & E_u & 0 & A_{1g} \end{pmatrix}$

3.3.4.2 Symmetry of the Diagonalization Matrix Elements

Let us now consider the symmetry of the transformation matrix elements in Table 3.1. The denominators N_1, N_2, N_3, and N_4 are invariant under all the D_{4h} operations. They contain groups of variables that follow the identical representation A_{1g}, as apparent in Table 3.3 or constants which are also invariant. Again by inspection of Table 3.3, the numerators of matrices $(T_{\gamma,A})$ and $(T_{\gamma,B})$ follow the irreducible representations presented in Table 3.7.

3.3.4.3 Permitted and Forbidden Optical Transitions

Let us now apply the group theory theorems described in Sect. 3.3.2 to the calculation of any of the envolvent terms of the optical dipole matrix element in Eq. (3.31). The operator involved will be x, y or z depending on the polarization. Their associated representations are (see Table 3.4) E_u, E_u or A_{2u} respectively. Each envelope function is calculated by (a) obtaining a diagonalized eigenfunction, whose associated representations are in Table 3.5 (in some cases the representation is the direct sum of two irreducible representations and therefore the corresponding character is the ordinary sum of characters), (b) obtaining its DFT, which is associated to the same representation (Sect. 3.3.3), (c) multiplying it by a $T_{\gamma,\boldsymbol{\kappa}}^{(u,v)}$ term, whose associated representation is again the direct sum of the representations of the terms $T_{\gamma A,\boldsymbol{\kappa}}^{(u,v)}$ and $T\gamma_{\alpha B,\boldsymbol{\kappa}}^{(u,v)}$, both appearing in Table 3.7, and finally (d) calculating the IDFT that also follows the same representation. This operation has to be done for the "initial" and "final" wavefunctions intervening in the dipole matrix element.

In summary, the sum (on symmetry operations, not on classes)

$$\sum_G \chi_G^{\Phi ini} \chi_G^{Tini} \chi_G^{M} \chi_G^{\Phi fin} \chi_G^{Tfin} \Big/ N_{\mathbf{G}} = h^{(A_{ig})} \tag{3.41}$$

is to be performed and checked if it is zero. The superscripts *ini* and *fin* refer respectively to the initial and final eigenfunctions intervening and M to the element of matrix, x, y or z. For characters bearing the superindex M and for many QN combinations associated with superindices Φ^{ini} and Φ^{fin}, the characters correspond to irreducible representations; in most T-matrix elements and in some QN combinations they are the direct sum of two irreducible representations. The calculation is very easy and can be performed by hand or programmed (even in a spreadsheet).

Table 3.8 Rules for permitted transitions concerning the parity of the initial and final state. The complementary set of the cases permitted in the table are forbidden transitions. Permitted transitions may still have a zero matrix element

Transition	Polarization	envelopes	Parity of initial and final states
$hh \rightarrow cb$ $so \rightarrow cb$	Horizontal (x,y)	all	• Different for one and only one of the x- and y-QNs • Irrelevant for the z-QN
	Vertical (z)	X,Y	• Either the same or different simultaneously for both x- and y-QNs • Irrelevant for the z-QN
		Z,S	• The same for both x-and y-QNs or the initial parity of the x-QN is the same as the final parity of the y-QN and vice versa • Irrelevant for the z-QN
$lh \rightarrow cb$	Horizontal (z,y)	X,Y	• Different for one and only one of the x- and y-QNs • The same for the z-QN
		Z,S	• Different for one and only one of the x- and y-QNs • Irrelevant for the z-QN
	Vertical (z)	X,Y	• Either the same or different simultaneously for both x- and y-QNs • Different for the z-QN
		Z,S	• The same for both x-and y-QNs or the initial parity or the x-QN is the same than the final parity of the y-QN and vice versa • Irrelevant for the z-QN

To have an idea of the calculation burden reduction, of the 1,728 different types of matrix elements when classified by symmetry, 976 of the $hh \rightarrow cb$ transitions and 1,180 of the $lh \rightarrow cb$ transitions are forbidden by symmetry reasons. For $\gamma = 0$, there are 1,524 forbidden $hh \rightarrow cb$ transitions and 1,352 forbidden $lh \rightarrow cb$ transitions. Some of the transitions permitted by symmetry may still have zero probability.

By examining all the symmetry cases, we can extract the rules in Table 3.8. Note that, for the case of $\gamma = 0$ and $\gamma = 1$, many or all of the reducible representations in the matrix $T_{\alpha,\kappa}^{(u,v)}$ become irreducible and the tables of forbidden elements are modified. Thus, Table 3.8 is not applicable to these cases. Note finally that, since the transitions $hh \rightarrow cb$ and $so \rightarrow cb$ follow the same rules, they apply to the standard and the interchanged configuration as well.

The interband absorption coefficients for transitions between bound states were considered in Sects. 3.2.3–3.2.5. Symmetry considerations were not taken into account and the zeroes in the matrix elements due to symmetry appeared as such by calculation (providing a solid check for our Table 3.8 of permitted transitions).

3.4 Interband Absorption of Photons by Extended States in Intermediate Band Solar Cells

In this section, absorption coefficients are calculated for transitions from the extended states in the valence band to the IB bound states. Symmetry considerations are rather important for this section. This completes the previous body of work in which transitions between bound states were calculated. The calculations are based on the EKPH considering the quantum dots as parallelepipeds with a very simple $U(\boldsymbol{r})$ function that is a squared potential well/pedestal of height V in all the QD volume. The extended states may be only partially extended—in one or two dimensions—or extended in all three dimensions. It is found that extended-to-bound state transitions are, in general, weaker than bound-to-bound state transitions, and that the former are weaker when the initial state is extended in more coordinates. This section is based on Ref. [34].

3.4.1 The Diagonalized Hamiltonian Solutions and the Energy Spectrum

As explained in Sect. 3.2.1, in the EKPH model, the calculation of the energy levels is easy. The effective mass equation is to be solved for each band with a potential energy that is the band offset due to the non-homogeneity of the material. In the VBs, the effective masses are negative and potential pedestals in the VBs behave as wells. These equations will give the spectrum of energies.

In the solution of the effective mass equation, a separation of variables approximation has been applied so that $\Phi(\boldsymbol{r}) = \xi(x)\psi(y)\zeta(z)$. Each one of these solutions is the solution of a one dimensional (1D) TISE in Eq. (2.38), V being the band offset and $2a$ the length of the QD in the x- and y-direction and $2c$ in the z-direction.

Finding the solutions for $\xi(x)$ [or for $\psi(y)$, $\zeta(z)$] constitutes a simple exercise of differential equations. Summarizing what is explained in Sect. 2.2, for $E_x < V$, bound solutions, different from the trivial $\xi(x) = 0$, are (dropping the subindex x) even ($\cos(kx)$) or odd ($\sin(kx)$) harmonic functions inside the well flanked by fading exponential functions outside it ($\exp(-\kappa x)\ \forall\, x \geq a$). Solutions may only exist for certain values of the wavenumbers k_n for which the non-fading exponential solution is cancelled. The energy E_n, fading coefficient κ_n and the wavevector values are related by Eq. (2.39). The index n denotes the different permitted energies in increasing order; this is a quantum number (QN). Odd QNs correspond to even functions and vice versa.

For $E \geq V$, the solution is harmonic, even or odd, with wavenumber k inside the potential well and also harmonic, even or odd, outside it but with a different value of the wavenumber k_e and a phase term. That is, they are of the form $\cos(k_e x - \theta)$ or $\sin(k_e x - \theta)$. Details can be found, e.g. in [22]. In this case, energy E_n, fading coefficient κ_n and the wavevector values are related by Eq. (2.41), θ being calculated through Eq. (2.42).

For $E \geq V$, k_e can take any value and therefore it leads to a continuum spectrum of energies. Since the mathematics of continuum spectra is rather complicated, it is common to assume that the wavefunctions are restricted to a large but finite region (a segment of length[3] $2L$, with large L, for one-dimensional cases, or a big parallelepiped for three-dimensional ones) and assume periodic conditions there. This leads [22] to the values in Eq. (2.43). Neglecting the variation of θ, the permitted values of k_e are separated by $\Delta k_e \cong \pi/2L$ (or $\Delta k_e \cong \pi/L$ for states of the same parity), which is small as long as L is big. However, only a numerable set of k_e-values are now permitted and the new QN, $\tilde{n}$, has appeared. Nevertheless, in most cases in this book, we will use k_e (with specification of parity), rather than $\tilde{n}$, for presentation purposes, although $\tilde{n}$ is used for calculations.

As already said, the three dimensional (3D) TISE solutions are labeled, within this approximation, as *cb*(111), *hh*(121) etc., the triplets of integers referring to the QNs of the 1D solutions. Their energy is calculated with Eq. (2.45). These energies, within this approximation, are represented in Fig. 3.4 for all the bands.

The validity of the separation of variables approximation has been examined in [35] and in Sect. 2.3. The energies obtained with the separation-of-variables approximation are very close to the exact ones for the bound states within the host material bandgap (IB states) and for the lowest virtual bound states (below 0.2 eV within the host CB) but after this the energy levels are pushed down, considerably for the highest energies. The first order approximation energies are shown for the relevant band (the *so* band is excluded) bound states in Fig. 3.5. They are rather close to the exact ones. In this section, we use this approximation (not yet used in Sect. 3.2), obtained by adding $\langle\Phi|U'|\Phi\rangle$ to the energy obtained in the zero order (separation of variables), where U' is the difference between the Hamiltonian and the one admitting as solutions those with separation of variables (see its expression in Sect. 2.3).

The energy of an absorbed photon is the difference between the final and initial electronic state energies. We call it E_{line}. In this case it is obtained by subtracting the VB state energy, calculated to the 1st order approximation, from the CB bound state energy within the bandgap, calculated in the separable approximation (further approximation makes no significant change). For a bound initial state it is,

$$E_{line} = E_G + E_{BScb} - V_{vb} + \left(\frac{\hbar^2(k_{nx}^2 + k_{ny}^2 + k_{nz}^2)}{2m^*}\right) + \langle\Phi|U'|\Phi\rangle \tag{3.42}$$

where E_{BScb} is the energy of the *cb* bound state, which is negative for the cases in study (as well as $\langle\Phi|U'|\Phi\rangle$) and is not further developed. Contrarily, we fully develop the energy of the initial VB bound state, which may belong to the *hh* or to the *lh* bands (*so* states have too low energy as to give E_{line} within the host bandgap).

[3] Note that in other instances within this book we call L what here is called $2L$.

The initial-state $\boldsymbol{k}$ components have discrete permitted values, and they are labeled with their correspondent nx, ny, nz quantum numbers.

When the wavefunction is extended in one of its variables, E_{line} is denoted in a more informative way: the subscript includes the extended variables outside the bracket and the QNs of the bound parts inside the bracket. According to this nomenclature, and taking into account the energy-wavevector relationship for extended states of Eq. (2.41), these equations are

$$\left\{ E_{z(nx,ny)} = E_G + E_{BScb} + \left(\frac{\hbar^2(k_{nx}^2 + k_{ny}^2)}{2m^*}\right) + \langle\Phi|U'|\Phi\rangle + \left(\frac{\hbar^2 k_{ez}^2}{2m^*}\right) \equiv E_{Tz(nx,ny)} + \left(\frac{\hbar^2 k_{ez}^2}{2m^*}\right) \right.$$
$$\left\{ \begin{array}{l} E_{xy(nz)} = E_G + E_{BScb} + V_{vb} + \left(\frac{\hbar^2 k_{nz}^2}{2m^*}\right) + \langle\Phi|U'|\Phi\rangle + \left(\frac{\hbar^2(k_{ex}^2 + k_{ey}^2)}{2m^*}\right) \\ \qquad \equiv E_{Txy(nz)} + \left(\frac{\hbar^2(k_{ex}^2 + k_{ey}^2)}{2m^*}\right) \end{array} \right.$$
$$\left\{ \begin{array}{l} E_{xyz} = E_G + E_{BScb} + 2V_{vb} + \langle\Phi|U'|\Phi\rangle + \left(\frac{\hbar^2(k_{ex}^2 + k_{ey}^2 + k_{ez}^2)}{2m^*}\right) \\ \quad = E_G + E_{BScb} + \left(\frac{\hbar^2(k_{ex}^2 + k_{ey}^2 + k_{ez}^2)}{2m^*}\right) \equiv E_{Txyz} + \left(\frac{\hbar^2(k_{ex}^2 + k_{ey}^2 + k_{ez}^2)}{2m^*}\right) \end{array} \right. \tag{3.43}$$

for 1E, 2E and 3E states respectively. The subindices clarify which variables are extended and which are bound; T stands for threshold. QNs always refer to bound variables. The formulas for other extended variables are obtained by permuting the variables. In all the cases, the continuous spectrum appears above a certain threshold that depends on the energy of the confined part of the eigenfunction. This is clearly visible in Fig. 2.6b where the lines in the regions 1E and 2E represent these thresholds. Above one of them there is a continuum of states with one or two k_e variables that can be freely chosen. However, what is reported in Fig. 2.6b corresponds to the CB, while in this section we are interested in the extended states in the VB (*lh* or *hh*).

Notice that, according Eq. (2.41), every extended dimension produced adds V_{vb} to the threshold energy, which is partly compensated by the negative value of $\langle\Phi|U'|\Phi\rangle$ so that the largest threshold energies approach the top of the VB [the smallest threshold energies approach to the bottom of the CB in Fig. 2.6b]. In the 3E case, $\langle\Phi|U'|\Phi\rangle = -2V_{vb}$, so that the two terms balance out. This means that the threshold for 3E functions in the VB is the VB top.

In this book, we are using a separable approximation for the wavefunctions. The use of exact values is not convenient because the wavefunctions are then linear combinations of separable wavefunctions (which is a basis), in most cases with one dominant basis function. This makes the labeling of the states cumbersome. Also, the absorption lines would be a mixture of two or more of the absorption coefficients used here, although with one dominating. Thus the use of the exact wavefunctions makes the calculations and their description inextricable. The use of the separable approximation for the wavefunctions (with a first-order perturbation correction for the energies) renders the problem manageable.

3.4.2 Calculation of the Absorption Coefficients

The absorption coefficient for transitions between levels under vertical (z) illumination can be found in Eq. (2.70). In Fig. 3.12, we present (dashed line) the sub-bandgap absorption for transitions between bound states (including the virtual bound states) under vertical illumination. Under the hypothesis of full collection of the generated electron-hole pairs, this absorption has permitted calculation of the internal quantum efficiency. It has agreed rather well with the measured value [11, 12]. This absorption has been calculated by calculating all the corresponding envolvents and employing Eqs. (2.70) and (3.29) of the absorption coefficient and of the dipole matrix element. For the calculation of the envolvents, the procedure, as described in Sect. 3.2.2 and in [11, 12], comprises four steps: (1) calculate the (initial and final) state eigenfunctions using the corresponding effective mass equation; (2) obtain their Fourier Transform; (3) multiply it by the appropriate element of the diagonalization matrix; (4) obtain its inverse Fourier Transform. The diagonalization matrix is a four dimensional square matrix whose rows correspond to the different bands (*cb, lh, hh, so*) and whose columns correspond to the possible projections ($|X\rangle$, $|Y\rangle$, $|Z\rangle$, $|S\rangle$). The Fourier Transforms are discrete Fourier Transforms, as is the usual case for numerical calculations. The total absorption is obtained as the sum of the absorption coefficients for all the lines between couples

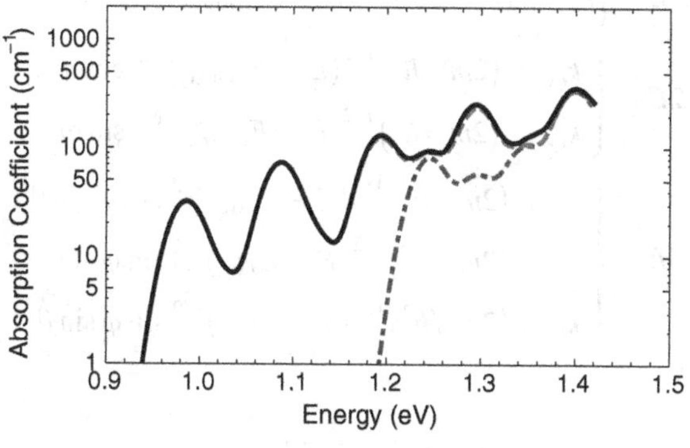

Fig. 3.12 *Dashed/red* total sub-bandgap absorption due to transitions between bound states. *Dot-dashed/Blue* sub-bandgap absorption between VB bound states and virtual bound states within the CB. *Solid/black* total sub-bandgap absorption due to the former transitions and transitions between bound and extended states. Calculations for $\gamma = 0.2$. Reproduced with permission. © 2013, Elsevier [34]

of bound states in the VBs and in the CB, limiting sometimes our calculations to the energies below the bandgap. The Dirac deltas appearing in Eq. (2.70) are drawn as Gaussian functions to account for the variations of level positions due e.g. to size variations of the QDs.

The same procedure is used in this section for the calculation of the transitions between the different types of extended states in the VB and the *cb* BS. However the continuous nature of the extended states implies substitution of the sums by integrals.

We now study the absorption coefficients for the transitions from 1E, 2E and 3E VB states to the *cb* BSs. We take as examples the transitions $hh(31e) \rightarrow cb(121)$, $lh(e2o) \rightarrow cb(111)$ and $hh(eoe) \rightarrow cb(221)$. The labels *e* and *o* refer to even or odd 1D extended states (not QNs). The expression for the 1E, 2E and 3E cases are respectively (see Eq. (2.70) for the definition of the α'),

$$\begin{aligned}
\alpha^{\max}_{hh(31e)cb(121)} &= \frac{L}{\pi}\int_0^\infty \alpha'_{hh(31e)cb(121)}(E)E\delta(E-E_{line})\frac{dk_{ez}}{dE}dE \\
\alpha^{\max}_{lh(e2o)cb(111)} &= \frac{L^2}{\pi^2}\int_0^{\pi/2} d\omega\int_0^\infty \alpha'_{lh(e2o)cb(111)}(E,\omega)E\delta(E-E_{line})\frac{d(k_{ex},k_{ez})}{d(E,\omega)}dE \\
\alpha^{\max}_{hh(eoe)cb(221)} &= \frac{L^3}{\pi^3}\int_0^{\pi/2} d\phi\int_0^{\pi/2} d\theta\int_0^\infty \alpha'_{hh(eoe)cb(221)}E\delta(E-E_{line})\frac{d(k_{ex},k_{ey},k_{ez})}{d(E,\theta,\varphi)}dE
\end{aligned} \tag{3.44}$$

where we have made the following changes of variables, respectively.

$$\begin{aligned}
&1E\quad \left\{k_{ez} = \left(2m^*/\hbar^2\right)^{1/2}\left(E-E_{Tz(nx,ny)}\right)^{1/2}\right. \\
&2E\quad \left\{\begin{array}{l} k_{ex} = \left(2m^*/\hbar^2\right)^{1/2}\left(E-E_{Txy(nx)}\right)^{1/2}\cos\omega \\ k_{ey} = \left(2m^*/\hbar^2\right)^{1/2}\left(E-E_{Txy(nx)}\right)^{1/2}\sin\omega \end{array}\right. \\
&3E\quad \left\{\begin{array}{l} k_x = \left(2m^*/\hbar^2\right)^{1/2}\left(E-E_{Txyz}\right)^{1/2}\cos\varphi\sin\theta \\ k_y = \left(2m^*/\hbar^2\right)^{1/2}\left(E-E_{Txyz}\right)^{1/2}\sin\varphi\sin\theta \\ k_z = \left(2m^*/\hbar^2\right)^{1/2}\left(E-E_{Txyz}\right)^{1/2}\sin\varphi\sin\theta \end{array}\right.
\end{aligned} \tag{3.45}$$

The Jacobians are,

$$\frac{dk_{ez}}{dE} = \left(\frac{m^*}{2\hbar^2(E - E_{Tz(nx,ny1)})}\right)^{1/2}$$
$$\frac{d(k_{ez}, k_{ey})}{d(E, \omega)} = \left(\frac{m^*}{\hbar^2}\right) \tag{3.46}$$
$$\frac{d(k_{ez}, k_{ey}, k_{ez})}{d(E, \theta, \varphi)} = \left(\frac{2^{1/3}m^*}{\hbar^2}\right)^{3/2} \sin\theta(E - E_{Txyz})^{1/2}$$

so that the absorption coefficients result as

$$\begin{cases} \alpha^{\max}_{hh(31e)cb(121)} = \dfrac{L}{\hbar\pi}\sqrt{\dfrac{m_{hh}}{2}}\dfrac{E\alpha'_{hh(31e)cb(121)}(E_{line})}{\sqrt{(E_{line} - E_{Tz(31)})}} \\ \alpha^{\max}_{lh(e2o)cb(111)} = \left(\dfrac{m_{lh}L^2}{\pi^2\hbar^2}\right)E_{line}\displaystyle\int_0^{\pi/2} \alpha'_{lh(e2o)cb(111)}(E,\omega)d\omega \\ \alpha^{\max}_{hh(eoe)cb(221)} = L^3\left(\dfrac{2^{1/3}m_{hh}}{\pi^2\hbar^2}\right)^{3/2}(E_{line} - E_{Txyz})^{1/2} \\ \qquad E_{line}\displaystyle\int_0^{\pi/2} d\varphi \int_0^{\pi/2} \alpha'_{hh(eoe)cb(221)} \sin\theta d\theta \end{cases} \tag{3.47}$$

Using the averaged values of the elementary-transition absorption coefficients (from now on E_{line} will be written E for brevity)

$$\hat{\alpha}'_{lh(e2o)cb(111)}(E) = \frac{2}{\pi}\int_0^{\pi/2} \alpha'_{lh(e2o)cb(111)}(E,\varphi)d\varphi$$
$$\hat{\alpha}'_{hh(eoe)cb(221)}(E) = \frac{2}{\pi}\int_0^{\pi/2} d\varphi \int_0^{\pi/2} \alpha'_{hh(eoe)cb(221)}(E,\theta,\varphi)\sin\theta d\theta \tag{3.48}$$

we can write

$$\alpha^{\max}_{hh(31e)cb(121)} = \frac{L}{\hbar\pi}\sqrt{\frac{m_{hh}}{2}}\frac{E\alpha'_{hh(31e)cb(121)}(E)}{\sqrt{(E - E_{Tz(31)})}}$$
$$\alpha^{\max}_{lh(e2o)cb(111)} = \left(\frac{L^2 m_{lh}}{2\hbar^2\pi}\right)E\hat{\alpha}'_{lh(e2o)cb(111)}(E) \tag{3.49}$$
$$\alpha^{\max}_{hh(eoe)cb(221)} = L^3\left(\frac{m^*}{2^{1/3}\hbar^2\pi^{4/3}}\right)^{3/2}(E - E_{Txyz})^{1/2}E\hat{\alpha}'_{hh(eoe)cb(221)}(E)$$

Notice that below the threshold energy no absorption is produced. For the 1E states, this absorption coefficient has a singularity (infinite) in $E_{Tz(12)}$, similar to that in the quantum wires. As done with the Dirac deltas for the bound states, here this singularity is smoothed with a soft step function.

To obtain all the absorptions from the 1Ez *hh* functions to the *cb*(111) bound state, we must add all the combinations of (n_x,n_y,e) and then (n_x,n_y,o) with n_x,n_y, running from 1 to 7. The symmetry rules in Sect. 3.3 (Table 3.8) may reduce the number of calculations (approximately divided by 8) because x and y are symmetric and the parity of either *nx* or *ny* has to differ once in the initial and final states [11]. To this we must sum the absorptions for the 1Ey where the quantum numbers are $(n_x,e/o,n_z)$ and the same with 1Ex.

Similar treatment has to be done with the *lh* band, adding again the resulting curves. In principle, the same should be done with the *so* band, but its energy position is too low and the transitions fall beyond the host bandgap, where we are not interested.

Similar sums have to be done with the 2E states and with the 3E states (in this latter case all the combinations of parities have to be summed up).

The global results are presented for all the 1E, 2E and 3E as initial states to the final states *cb*(111), *cb*(121) [including the degenerate *cb*(211)] and *cb*(221) in Fig. 3.13. In this calculation, the IB has been considered empty so presenting the highest absorption. The (long-dashed) blue curves represent the transitions between bound states, as calculated in [11], (medium-dashed) red, (short-dashed) green and (dotted) purple curves correspond to the 1E, 2E and 3E initial states. It can be seen that the absorptions from the 1E states tend to be dominant for photons close to the bandgap energy. Furthermore, 2E absorptions are smaller and 3E absorptions are negligible. The onset of the absorption coefficient is located in all cases near the VB top. In the 3E case this is strictly at the VB top.

3.5 Discussion

The most straightforward conclusion of this section is represented in Fig. 3.13. It can be concluded that the extended states absorb less light than the bound states, excepting a range of energies close to the bandgap. However, in this region, they are dominant and would amend the relatively lower increase of the calculated (only with bound states) quantum efficiency observed in Fig. 3.10 with respect to the measured one; also the curve becomes less wavy. This is intuitive because the occurrence of a transition requires an overlap of the initial and final states. When one of these is extended, then the overlap is reduced; this is partly compensated by the existence of many extended sates, but not sufficiently to give strong absorption. Nevertheless, for high energy photons the absorption between bound states is negligible and the absorption from 1E states becomes dominant. 1E extended states lead to higher absorption than 2E and much more than 3E but still smaller than 0E states. This behavior is to be compared to the transitions IB → extended CB states

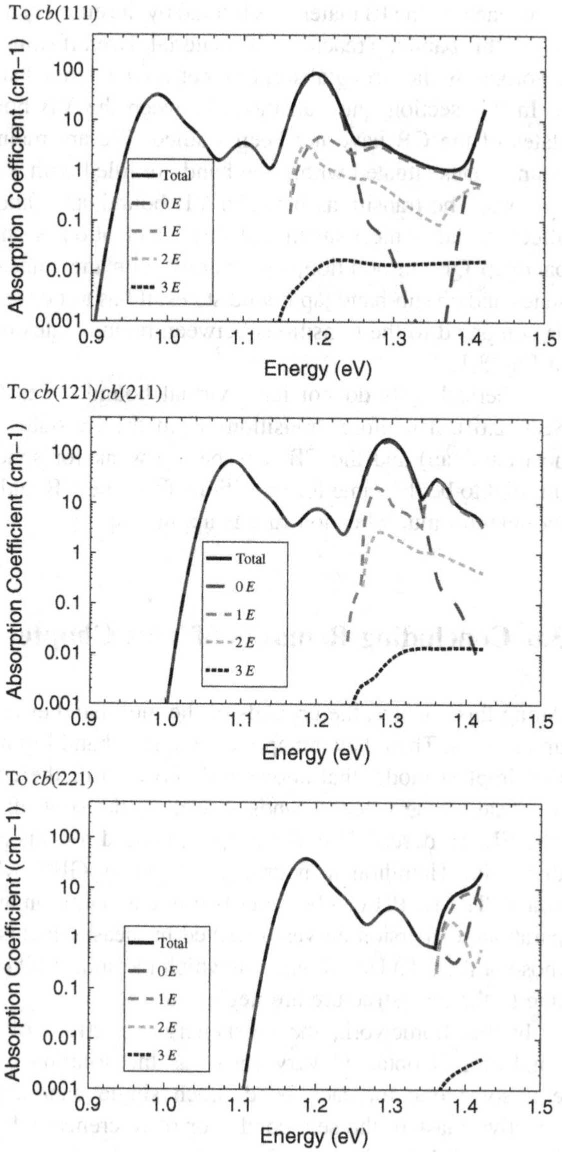

Fig. 3.13 Absorption coefficient for the transitions from the VBs and the bound states of the CB. *Blue* form the 0E bound states. *Red* form the 1E states. *Green* form the 2E states. *Purple* form the 3E states. *Black* sum of all the preceding. Reproduced with permission. © 2013, Elsevier [34]

in Sect. 2.4.4, where transitions to 2E and 3E states are strictly forbidden (under the separable approximation).

The curves in Fig. 3.13 provide data for an exact detailed balance calculation of the intermediate band solar cell. However, neglecting the contribution of the extended states may be a reasonable simplifying option.

In this section, only absorptions within the host bandgap range of energies are calculated. The reason for not considering photons beyond the bandgap is that the light reaching the IB material is filtered by the emitter so that few photons with energy above this bandgap reach the IB material. Even if some reach it, they will be mainly absorbed by the strong absorption between extended states of the VB and the CB.

In this section, the transitions between the VB bound states and the extended states of the CB have not been studied. We are mainly interested in how the *cb* bound states situated within the band are filled with VB electrons by absorption of photons. The transitions between VB bound sates and CB extended states do not affect the latter mechanism and their calculation is only useful to predict the sub-bandgap QE. In addition, by analogy with the transitions between VB extended states and *cb* sub-bandgap bound states, it has to be presumed that their role is small as compared to the transitions between bound states represented by the blue curve in Fig. 3.12.

Spherical QDs do not have virtual bound states (see e.g. [30, Chap. 11 and Sect. 2.6]). Therefore, transitions from the IB states (cb bound states within the host bandgap) and the CB may be too weak for spherical QDs. This behavior is inferred to be the same for transitions form the VB and the IB states, at least, whilst the integral multiplication rule is applicable.

3.6 Concluding Remarks of This Chapter

Along this chapter, the grounds of the multiband envelope equations are described and proven. Then they are applied to a four band k·p model, which is thought to be the simplest model that allows a description of the conduction band and the light and heavy hole valence bands as well as the split off band in zincblende crystals. The EKPH, described in the chapter, is used for this purpose. This involves a four dimension Hamiltonian matrix generated by GBF, which are doubly spin degenerated. The EKPH can be described like a Hamiltonian whose eigenvectors are the parabolic dispersion curves observed in measurements and whose eigenvectors are those of the (H_0) Hamiltonian in which the spin-orbit coupling and the strain effects due to the nanostructure are neglected.

In this framework, the eigenenergy spectrum of a QD nanostructured semiconductor is obtained very easily as the solution of four single band envelope equations: one for each band. Each single band equation has a corresponding effective mass in the semiconductor of reference, which we recommend to be the QD material for the bound functions and the host semiconductor for extended functions. In this chapter, this has been applied to the SOTA solar cell under the assumption of a box shaped potential well. Although not attempted, the Bastard boundary conditions might lead to more accurate calculations when bound and extended states are simultaneously involved in the calculations.

The use of the experimental effective masses for each band is a basic assumption of the EKHP method and is the way of introducing the spin orbit interaction and the strain effects of the nanostructure. The rest of the assumptions in the chapter—the square potential wells, the box shaped form of the QD, and the separation-of-variables methods—are contingent, but they are very important to permit quick calculations with reasonable results.

Since (H_0)—without spin-orbit coupling or strain effects—has a doubly degenerate eigenvalue, the corresponding eigenvectors [that form (T)] can be chosen amongst the infinite possibilities within the resulting vector space. The first choice made is a pair of orthonormal eigenvectors that transform in the simplest way under the symmetry group of the Hamiltonian. In this choice, each of their elements follows an irreducible representation of the D_{4h} group. The remaining possibilities are then sampled by Givens-rotating this pair by a rotation parameter γ—where the rotation axis is defined so as to preserve the symmetry between the *X*- and *Y*-GBFs—and by swapping the eigenvectors that corresponds to the *hh* and to the *so* band. Once rotated, the matrix elements usually follow reducible representations in the group D_{4h}.

Absorption spectra are presented [11] for all possible cases and the role of γ is discussed. The agreement of the calculated absorption spectrum using the model in [11] is found to be better than that using the previous model, presented in [12], in which a detailed study of symmetry was absent. The best agreement with experimental results occurs for small γ.

It is found that the weakness of the absorption is mostly due to the fact that the CB confined states have a small projection onto the VB *X*- and *Y*-GBFs and that the VB confined states have a small projection onto the CB *S*-GBF. How to increase the overlap between VB and CB states by design of the QD structure remains an open question.

The sub-bandgap absorption by transitions between VB and CB bound states is represented in Fig. 3.12 (dashed curve). When the absorption between VB extended sates and the *cb* BS are added, the black curve is obtained. It is apparent that the extended states affect the absorption curve very slightly and that our assumption in [11, 12], that only the transitions between bound states is important, is rather accurate. The agreement between the calculated and the measured QE, that was a justification of the model, is confirmed by the results of Sect. 3.4.

There are important transitions linking directly the bound states of the VB to the virtual bound states within the CB. These are represented in Fig. 3.9 for QDs of several sizes and in Fig. 3.12 by the dot-dashed curve (blue in the electronic version) in the SOTA prototype. Actually, this constitutes a shrinking of the optical bandgap due to the QDs and in a wider sense reflects transitions between the VB and the CB. This is so because the *hh* states are in thermal contact with the whole VB due to the small energy interval between them. The same can be said about the *lh* states, which are immersed in the *hh* range of energies. On the other hand, the *cb* VBS are within the CB and therefore in thermal equilibrium with it. This shrinkage is visible in Fig. 3.4 where the *hh* states represent the extent to which the VB enters into the GaAs bandgap.

This bandgap shrinkage does not necessarily imply a substantial reduction of the voltage of the solar cells currently manufactured with this material, as shown in [36] and [37]. In reality, the density of (*hh* and *lh*) states introduced within the band by the QDs is rather faint (as confirmed by the weak absorption of light). The situation may change if a much bigger density of QDs is introduced. This topic will be further developed in Chap. 4.

Observe also in Fig. 3.12 that the influence of the extended states is really important beyond 1.38 eV. This is the region in which a rather strong absorption is produced, attributed to the wetting layer associated to the manufacturing of the QDs by growth of epitaxial layers of InAs on GaAs in the Stranski Krastanov mode. This layer actually forms a quantum well whose absorption is much stronger. No attempt to model the QE in this region has been made in [11, 12].

From a physical point of view, we think that a rather clear picture of the sub-bandgap absorption process is now available. The IB *cb*(1,1,1) confined state is partially filled of electrons due to n-doping of the CB and receives electrons pumped by photons from the *hh* states that form a deep VB queue inside the host bandgap (see Fig. 3.4) and also form the *lh* and *so* bands (the latter with negligible influence because it requires overly energetic photons), but this is only a small part of the sub-bandgap photocurrent. Many of the photons are absorbed by transitions to the remaining *cb* confined states, and in particular by *cb*(1,2,1), *cb*(2,2,1) and *cb* (1,3,1) (see Fig. 3.9 and Fig. 3.10) originated in the *hh* and *lh* bands. Numerous transitions take place among these cb states [21] associated to absorptions and emissions of low energy photons (of about 100 meV) taken from the huge reservoir of thermal photons within the semiconductor and to a lesser extent from photons coming from outside. Finally, they end in confined states of larger QNs embedded in the CB and from there they pass to the external circuit. For instance, in the SOTA solar cell, the easier path for the IB states *cb*(1,1,1) to the CB is [21] *cb*(1,1,1) → *cb* (1,2,1) → *cb*(1,3,1).

References

1. Dresselhaus G, Kip AF, Kittel C (1955) Plasma resonance in crystals—observations and theory. Phys Rev 100(2): 618–625. doi:10.1103/PhysRev.100.618
2. Kane EO (1956) Energy band structure in p-type germanium and silicon. J Phys Chem Solids 1(1–2):82–99. doi:10.1016/0022-3697(56)90014-2
3. Kane EO (1966) The k·p method. In: Willardson RK, Beer AC (eds) Physics of III-V compounds. semiconductors and semimetals, vol 1. pp 75–100. Academic, New York
4. Bastard G (1990) Wave mechanics applied to semiconductor nanostructures. Monographies de Physique. Les Editions de Physique, Paris
5. Basu PK (1997) Theory of optical processes in semiconductors. Series on semiconductor science and technology. Oxford University Press, Oxford
6. Wang LW, Zunger A (1994) Electronic-structure pseudopotential calculations of large (approximate-to-1000 atoms) Si quantum dots. J Phys Chem 98(8):2158–2165. doi:10.1021/j100059a032
7. Ivchenko EL (2005) Optical spectroscopy of semiconductors nanostructures alpha science, Harrow, UK
8. Voon LCLY, Willatzen M (2009) The k·p method. Electronic properties of semiconductors. Springer, Berlin
9. Galeriu C (2005) K·P theory of semiconductor nasnostructures. PhD disertations, Worcester Polytechnic Institute, Worcester

10. Datta S (1989) Quantum phenomena. Molecular series on solid state devices, vol 8. Addison Wesley, Reading (Mass), Boston
11. Luque A, Mellor A, Antolin E, Linares PG, Ramiro I, Tobias I, Marti A (2012) Symmetry considerations in the empirical k·p hamiltonian for the study of intermediate band solar cells. Sol Energy Mater Sol Cells 103:171–183
12. Luque A, Marti A, Antolín E, Linares PG, Tobías I, Ramiro I, Hernandez E (2011) New hamiltonian for a better understanding of the quantum dot intermediate band solar cells. Sol Energy Mater Sol Cells 95:2095–2101. doi:10.1016/j.solmat.2011.02.028
13. Mellor A, Luque A, Tobías I, Martí A (2013) Realistic detailed balance study of the quantum efficiency of quantum dot solar cells. Adv Funct Mater. doi:10.1002/adfm.201301513
14. Popescu V, Bester G, Zunger A (2009) Coexistence and coupling of zero-dimensional, two-dimensional, and continuum resonances in nanostructures. Phys Rev B 80(4):045327. doi:10.1103/PhysRevB.80.045327
15. Pikus GE, Bir GL (1959) Effect of deformation on the energy spectrum and the electrical properties of imperfect germanium and silicon. Soviet Physics-Solid State 1(1):136–138
16. Pikus GE, Bir GL (1961) Cyclotron and paramagnetic resonance in strained crystals. Phys Rev Lett 6(1): 103. doi:10.1103/PhysRevLett.6.103
17. Pryor C (1998) Eight-band calculations of strained InAs/GaAs quantum dots compared with one-, four-, and six-band approximations. Phys Rev B 57(12):7190–7195
18. Luque A, Linares PG, Mellor A, Andreev V, Marti A (2013) Some advantages of intermediate band solar cells based on type II quantum dots. Appl Phys Lett 103:123901. doi:10.1063/1.4821580
19. Linares PG, Marti A, Antolin E, Luque A (2011) III-V Compound semiconductor screening for implementing quantum dot intermediate band solar cells. J Appl Phys 109:014313
20. Sze SM (1981) Physics of semiconductor devices. Wiley, New York
21. Luque A, Marti A, Antolín E, Linares PG, Tobias I, Ramiro I (2011) Radiative thermal escape in intermediate band solar cells. AIP Adv 1:022125
22. Luque A, Marti A, Mellor A, Marron DF, Tobias I, Antolín E (2012) Absorption coefficient for the intraband transitions in quantum dot materials. Prog Photovoltaics. doi:10.1002/pip.1250
23. Mellor A, Luque A, Tobias I, Marti A (2013) A numerical study into the influence of quantum dot size on the sub-bandgap interband photocurrent in intermediate band solar cells. AIP Adv 3:022116
24. Mellor A, Luque A, Tobias I, Marti A (2012) The influence of quantum dot size on the sub-bandgap intraband photocurrent in intermediate band solar cells. Appl Phys Lett 101:133909. doi:10.1063/1.4755782
25. Antolín E, Marti A, Farmer CD, Linares PG, Hernández E, Sánchez AM, Ben T, Molina SI, Stanley CR, Luque A (2010) Reducing carrier escape in the InAs/GaAs quantum dot intermediate band solar cell. J Appl Phys 108(6):064513
26. Oshima R, Hashimoto T, Shigekawa H, Okada Y (2006) Multiple stacking of self-assembled InAs quantum dots embedded by GaNAs strain compensating layers. J Appl Phys 100(8):083110. doi:10.1063/1.2359623
27. Oshima R, Akahane K, Tsuchiya M, Shigekawa H, Okada Y (2007) Optical properties of stacked InAs self-organized quantum dots on InP (311)B. J Cryst Growth 301:776–780. doi:10.1016/j.jcrysgro.2006.11.063
28. Marti A, Antolin E, Canovas E, Lopez N, Linares PG, Luque A, Stanley CR, Farmer CD (2008) Elements of the design and analysis band solar of quantum-dot intermediate cells. Thin Solid Films 516(20):6716–6722. doi:10.1016/j.tsf.2007.12.064
29. Mellor A, Tobías I, Martí A, Luque A (2011) A numerical study of Bi-periodic binary diffraction gratings for solar cell applications. Sol Energy Mater Sol Cells 95(12):3527–3535. doi:10.1016/j.solmat.2011.08.017
30. Messiah A (1960) Mécanique Quantique. Dunod, Paris
31. Iniguez-Almech J-M (1949) Mecanica Cuantica. Memorias, Serie 2, Memoria 2, vol Serie 2, Memoria 2. Academia de Ciencias de Zaragoza

32. Atkins WP, Child MS, Phillips CSG (2006) Tables for group theory. Oxford University Press, Oxford
33. Kaxiras E (2003) Atomic and electronic structure of solids. Cambridge University Press, New York
34. Luque A, Mellor A, Ramiro I, Antolín E, Tobías I, Martí A (2013) Interband absorption of photons by extended states in intermediate band solar cells. Sol Energy Mater Sol Cells 115:138–144
35. Luque A, Mellor A, Tobías I, Antolín E, Linares PG, Ramiro I, Martí A (2013) Virtual-bound, filamentary and layered states in a box-shaped quantum dot of square potential form the exact numerical solution of the effective mass Schrödinger equation. Phys B 413:73–81. doi:10.1016/j.physb.2012.12.047
36. Tobías I, Luque A, Martí A (2011) Numerical modeling of intermediate band solar cells. Semicond Sci Technol 26:014031
37. Tobías I, Luque A, Antolín E, Linares PG, Ramiro I, Hernández E, Martí A (2012) Realistic performance prediction in nanostructured solar cells. J Appl Phys 112:24518. doi:10.1063/1.4770464

Chapter 4
Detailed Balance Analysis

Abstract In this chapter, a detailed balance model based on realistic absorption assumptions is presented for the study of QD-IBSCs in the radiative limit. This model helps us to analyse existing QD-IBSCs, and to make predictions about future proposals for QD-IBSC prototypes, both of which are done in this chapter. Sub-bandgap photon absorption/emission depends not only on the optical matrix elements, but on the electron occupation of the different bands and levels in the system. Neglecting non-radiative processes, these can be calculated by considering the so-called detailed balance of photon absorption and emission for each electronic transition and exacting the continuity of generation and recombination currents via each IB level. In this chapter, the model has been applied to calculation of the internal quantum efficiency (IQE) of a real QD-IBSC and its temperature dependence. The results are compared to previously published experimental data, with good agreement. The good agreement with experimental results makes this model a useful tool for making a realistic assessment of the feasibility of achieving high efficiencies using InAs/GaAs QD arrays, and to quantify the advances that are required to do so. Such an assessment is made at the end of the chapter.

Keywords Solar cells · Quatum dots · Photon absorption · Photon emission · Radiative detailed balance

Detailed balance considerations have been used in the past to calculate the efficiency limits of IBSCs [1] and single bandgap solar cells [2], but in these cases, full absorption of photons in each transition was assumed. In the model presented in this chapter, we do not assume full absorption, but instead take the optical matrix elements calculated in Chaps. 2 and 3 as input for the detailed balance model. This allows a realistic estimate of the sub-bandgap current generated by the QD-IBSC in the radiative limit.

In this chapter, the model has been applied to calculation of the internal quantum efficiency (IQE) of a real QD-IBSC and its temperature dependence. The results are compared to previously published experimental data [3, 4], with good agreement. Historically, this experiment has served two main purposes. It confirmed that the

A. Luque and A.V. Mellor, *Photon Absorption Models in Nanostructured Semiconductor Solar Cells and Devices*, SpringerBriefs in Applied Sciences and Technology, DOI 10.1007/978-3-319-14538-9_4

sub-bandgap photocurrent is too low, and established the thermal escape of electrons from the IB to the CB as a cause of the reduction of V_{oc} with respect to an equivalent reference cell with no QDs.

Using the model, we can analyse the individual sub-bandgap transition currents that make up the measured internal quantum efficiency (IQE), something that is not possible experimentally. We confirm that the poor sub-bandgap photocurrent is a consequence of the weak photon absorption investigated in previous works [5–9]. We also confirm that the thermal escape can be explained by radiative processes involving thermal photons at room temperature.

4.1 Energy Levels and Bands in the Exemplary QD-IBSC

As in the previous chapters, the modelled QD-IBSC is the here-denoted SOTA cell, which is the sample labelled SB in Ref. [3] and S3 in Ref. [4]. In this chapter we largely follow [10]. A simplified band diagram of a single QD in this exemplary QD-IBSC is shown in Fig. 4.1. The grey lines represent the VB and CB band edges, the offsets being due to the InAs QD.

The confining potential in the CB supports three discrete bound state levels in the energy range of the host forbidden band. Following the nomenclature from the

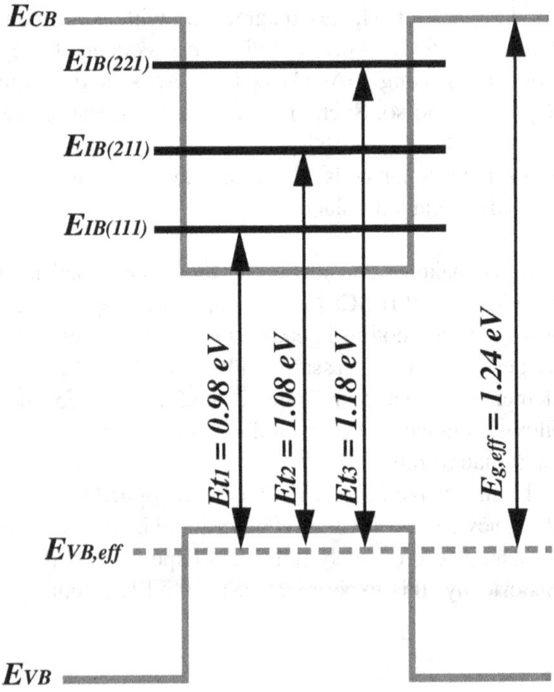

Fig. 4.1 A simplified band diagram of a single QD in the exemplary QD-IBSC. *Upper grey line* conduction band edge. *Lower grey line* valence band edge. *Black lines* confined state energy levels whose energy is within the host forbidden band. *Dashed grey line* effective valence band edge

previous chapters, these are labelled IB(111), IB(211) and IB(221).[1] The IB(211) level corresponds to the pair of states IB(211)/IB(121), which are degenerate due to the QDs' reflection symmetry in the $x = y$ plane. The energy of each level is denoted $E_{IB(111)}$, $E_{IB(211)}$ and $E_{IB(221)}$. It is not a priori assumed that the IB levels are thermally coupled and as such their electron population is described by Fermi-Dirac functions with three distinct quasi Fermi levels (QFLs): $E_{F,IB(111)}$, $E_{F,IB(211)}$ and $E_{F,IB(221)}$. The onset of the host CB is denoted E_{CB}. Above this energy, there is a continuum of extended states and a set of discrete virtual bound states [11] (neither pictured), all of which are described by a single QFL denoted $E_{F,CB}$.

The confining potential pedestal in the VB supports numerous bound heavy-hole and light-hole states, (not pictured). The heavy-hole states are so close together as to form a pseudo continuum, and it is therefore assumed that all confined hole states are thermally coupled to one another and to the rest of the VB. The VB is therefore considered to have its energy onset at the confined heavy-hole ground state energy, which is near the top of the pedestal ($E_{VB,\mathrm{eff}}$—dashed line). The Fermi-Dirac distribution of the VB is described by a single QFL, denoted $E_{F,VB}$. The lower density of states in the VB pedestal compared to the matrix VB is taken into account. The energies of the principal transitions between the VB pedestal and the IB levels, E_{t1}, E_{t2}, E_{t3}, are given in the figure.

4.2 The Detailed Balance Model

The steady state subbandgap currents in the exemplary QD-IBSC are described by five quasi Fermi levels (QFLs) one for each carrier population. These are found by solution of five simultaneous equations. There are three current continuity equations, one for each IB level, which are derived in Sect. 4.2.1. There is also a charge conservation condition, and an expression relating the VB-CB QFL split to the external bias voltage, these are described in Sects. 4.2.2 and 4.2.3 respectively. The solution of the problem to yield the output current is given in Sect. 4.2.4.

Since each IB level presents one unknown QFL and provides one continuity equation, the model can be extended to a QD system with any number of IB levels. The bias voltage, cell temperature, illumination conditions and optical matrix elements are the input parameters to the model. Hence the model can be applied to many QD-IBSC systems under different conditions.

4.2.1 Continuity of the Subbandgap Currents

In the exemplary QD-IBSC, we have five distinct electron populations: that of the CB, that of the VB (whose upper energy onset is at $E_{VB,eff}$), and those of the three

[1] In previous chapters these states were usually denoted *cb*(111), *cb*(211) and *cb*(221).

discrete IB energy levels. Generation and recombination currents exist between the five populations due to the respective emission and absorption of photons. The net generation current (generation or upward traffic of electrons up minus recombination or downward traffic) between a given pair of bands/levels is denoted $J_{l\to u}$, where l and u are respectively the lower and upper level or band. Due to the conventional GaAs layers on either side of the QD stack, current can only be extracted via the VB and CB and not directly from the IB levels. Hence, for each IB level with index k, we can write a current continuity equation:

$$\sum_{l} J_{l\to k} - \sum_{u} J_{k\to u} = 0 \tag{4.1}$$

where the first and second sums are made over all levels (or bands) whose energy is respectively lower or higher than that of IB level k. Equation (4.1) presents three equations; one for each intermediate band level. These are the first three of the set of five equations from which the QFLs are calculated. In order to solve them, we must express each $J_{l\to u}$ as a function of known parameters and the unknown QFLs.

The IB material is modelled as a homogeneous slab of thickness W in the z direction and extending infinitely in the xy plane. The IB layer is sandwiched between a semi-infinite GaAs substrate at the rear and a thin GaAs emitter at the front. The thin GaAs emitter screens all photons with $E_{phot} > E_g$ from reaching the IB layer, but interacts negligibly with photons with $E_{phot} < E_g$. All materials in the structure have the same refractive index n_{ref}, and the medium in front of the emitter is air with a refractive index of 1. Perfect transmission is assumed at all interfaces inside the semiconductor.

Both the air at the front and the GaAs substrate at the rear are a source of ambient photons at the cell temperature T_c and at zero chemical potential (we are calling ambient temperature to the cell temperature). From the front, these photons enter the IB layer through the escape cone of critical angle θ_c, given by Snell's law. From the rear, these photons enter the IB layer through the whole hemisphere, given that the IB layer and substrate are of the same refractive index. The structure is also illuminated from the front by an illuminating source (for example a monochromator or the sun).

We consider a single photon mode with wavevector $\boldsymbol{q}$ and unit polarization vector ε inside the IB layer, and denote this mode $\boldsymbol{q}$, ε. Note that, unlike in previous chapters, this definition of a mode is polarization specific. In the steady state, the continuity equation for this photon mode is

$$\frac{dN_{ph,\boldsymbol{q}\varepsilon}}{dz} = -\frac{n_{ref}}{c}\frac{1}{\cos\theta_q}\dot{N}_{ph,\boldsymbol{q}\varepsilon} \tag{4.2}$$

where $N_{ph,\boldsymbol{q}\varepsilon}$ is the number of photons in the mode, c is the speed of light in a vacuum, θ_q is the polar angle made between $\boldsymbol{q}$ and the z axis and $\dot{N}_{ph,\boldsymbol{q}\varepsilon}$ is the net annihilation of photons in the mode due to all possible absorption and emission processes per unit time.

For $E_{phot} < E_g$, we assume that the dominant absorption and emission processes are electronic transitions involving the IB levels. We consider a transition between respective upper and lower quantum states $|\Xi_u\rangle$ and $|\Xi_l\rangle$ with energy levels E_u and E_l and quasi Fermi levels $E_{F,u}$ and $E_{F,l}$. The probability that the states $|\Xi_u\rangle$ and $|\Xi_l\rangle$ are occupied by an *electron* is respectively given by the Fermi-Dirac functions f_u and f_l, where

$$f_x = \left[\exp\left(\frac{E_x - E_{F,x}}{k_B T_c}\right) + 1\right]^{-1} \tag{4.3}$$

T_c being the temperature of the solar cell.

For a single QD, the Fermi Golden Rule [12] states that the probability of an electron in quantum state $|\Xi_l\rangle$ making a transition to state $|\Xi_u\rangle$ due to emission/absorption of a photon in mode $\boldsymbol{q}$, $\boldsymbol{\varepsilon}$ is

$$\begin{aligned} w^a_{\boldsymbol{q},\epsilon} &= N_{ph,\boldsymbol{q}\epsilon} \frac{2\pi^2 e^2 E_{phot}}{\Omega n^2_{ref}\varepsilon_0} |\langle \Xi_l|\boldsymbol{r}\cdot\epsilon|\Xi_u\rangle|^2 \delta(E_u - E_l - E_{phot}) \\ w^e_{\boldsymbol{q},\epsilon,} &= (N_{ph,\boldsymbol{q}\epsilon} + 1) \frac{2\pi^2 e^2 E_{phot}}{\Omega n^2_{ref}\varepsilon_0} |\langle \Xi_l|\boldsymbol{r}\cdot\epsilon|\Xi_u\rangle|^2 \delta(E_u - E_l - E_{phot}) \end{aligned} \tag{4.4}$$

where $|\langle \Xi_l|\boldsymbol{r}\cdot\boldsymbol{\varepsilon}|\Xi_u\rangle|$ is the optical matrix element for the transition between the two states, discussed and calculated in Chaps. 2 and 3, and Ω is the crystal volume. Actually, Eq. (4.4) completes Eq. 2.67 of the Chap. 2. The absorption (emission) can only occur if there is an electron in the lower(upper) state and no electron in the upper(lower) state. The number of absorption and emission events per unit volume and time is therefore

$$\begin{aligned} a_{\boldsymbol{q},\boldsymbol{\varepsilon}} &= \rho_{QD} w^a_{\boldsymbol{q},\boldsymbol{\varepsilon}} f_l(1 - f_u) \equiv \frac{c}{n_{ref}} \frac{N_{ph,\boldsymbol{q}\boldsymbol{\varepsilon}}}{\Omega} \alpha^{\max}_{\boldsymbol{q},\boldsymbol{\varepsilon}} f_l(1 - f_u) \\ e_{\boldsymbol{q},\boldsymbol{\varepsilon}} &= \rho_{QD} w^e_{\boldsymbol{q},\boldsymbol{\varepsilon}} f_u(1 - f_l) \equiv \frac{c}{n_{ref}} \frac{(N_{ph,\boldsymbol{q}\boldsymbol{\varepsilon}} + 1)}{\Omega} \alpha^{\max}_{\boldsymbol{q},\boldsymbol{\varepsilon}} f_u(1 - f_l) \end{aligned} \tag{4.5}$$

where ρ_{QD} is the number of QDs per unit volume and $\alpha^{\max}_{\boldsymbol{q},\boldsymbol{\varepsilon}}$ is defined in Eq. (4.5) for convenience. The choice of this nomenclature was justified in Eq. (2.70) and is also explained shortly later.

In general, a single photon mode can couple with transitions between different pairs of states. However, in the weak absorption approximation, the photon population varies only slightly throughout the IB layer. We would therefore expect that the final transition rates between each pair of states to be negligibly affected by the photons coupling with competing transitions. We can therefore simplify the mathematics by considering each pair of states to couple exclusively to its own set of photon modes. Under this approximation, the net annihilation of photons in mode $\boldsymbol{q}$, $\boldsymbol{\varepsilon}$ due to transitions between $|\Xi_u\rangle$ and $|\Xi_l\rangle$ is

$$\dot{N}_{ph,\boldsymbol{q\varepsilon}} = \Omega(a_{\boldsymbol{q},\varepsilon} - e_{\boldsymbol{q},\varepsilon}) \tag{4.6}$$

and Eq. (4.2) becomes

$$\frac{dN_{ph,\boldsymbol{q\varepsilon}}}{dz} = \frac{1}{\cos\theta_q}\alpha_{q,\varepsilon}^{\max}(f_l - f_u)\left[N_{ph,\boldsymbol{q\varepsilon}} + \frac{f_u(1-f_l)}{f_u - f_l}\right] \tag{4.7}$$

We observe that

$$\begin{aligned}\frac{f_u(1-f_l)}{f_u - f_l} &= -\left[\exp\left(\frac{E - (E_{F,u} - E_{F,l})}{k_B T_c}\right) - 1\right]^{-1} = -\left[\exp\left(\frac{E-\mu}{k_B T_c}\right) - 1\right]^{-1} \\ &\equiv -f_{B,\mu T_c}\end{aligned} \tag{4.8}$$

$f_{B,\mu Tc}$ being the Bose-Einstein energy distribution of luminescent photons at temperature T_c and chemical potential $\mu = E_{F,u} - E_{F,l}$. The general solution to Eq. (4.7) is

$$\begin{aligned}N_{ph,\boldsymbol{q\varepsilon}}(z) &= A\exp\left\{-\alpha_{q,\varepsilon}^{\max}(f_l - f_u)\frac{z}{\cos\theta_q}\right\} + f_{B,\mu T_c} \\ &\cong A\left(1 - \alpha_{q,\varepsilon}^{\max}(f_l - f_u)\frac{z}{\cos\theta_q}\right) + f_{B,\mu T_c}\end{aligned} \tag{4.9}$$

with A the constant of integration. In the last step, the exponential has been approximated as first order Taylor expansion, which is valid for weak absorption. For a single highly oblique order, $\cos\theta_q$ is very small and the first order approximation is not valid; however, preliminary calculations have found that the error incurred when integrating over all angles is small for the absorption coefficients considered in this chapter.

To find the constant of integration in Eq. (4.9) we apply boundary conditions determined by the solar cell geometry and operating conditions. The solar cell is illuminated from the front by photons travelling in the positive z direction. At the front surface, photons enter and escape the IB layer within the escape cone determined by n_{ref}. At the rear surface, photons enter and escape the IB layer over the whole hemisphere. In the following, we distinguish between modes inside the escape cone of the front surface (superscript "esc") and those outside the escape cone (they are confined by total internal reflection, superscript "conf"), and between modes propagating in the positive z direction (superscript "+") and those travelling in the negative z direction (superscript "–").

Modes within the escape cone travelling in the positive z direction enter the IB material at $z = 0$. Their population is equal to the population incident from the illuminating source (after being filtered by the emitter), which is denoted $N_{inc}(\theta_q, \varphi_q)$ and in a given direction is assumed to be the same for all polarizations (unpolarized incident beam), plus the photons incident from the surroundings at room temperature. Modes travelling in the negative z direction enter the IB material at $z = W$ from the GaAs substrate. It is reasonable to think that sub-bandgap photons entering from the doped substrate have zero chemical potential under any voltage conditions. This is because they do not have sufficient energy to interact with the non-equilibrium VB-CB transitions, and instead interact with electrons or holes in a single band, or phonons, the individual populations of which are assumed to be in equilibrium. Finally, modes outside the escape cone travelling in the positive z direction are those that enter from the rear and are reflected (by total internal reflection) at $z = 0$. The boundary conditions for each class of mode are therefore

$$\begin{aligned} N_{ph,q\varepsilon}^{+esc}(0) &= N_{inc}(\theta_q, \varphi_q) + f_{B,0T_c} \\ N_{ph,q\varepsilon}^{-esc}(W) &= N_{ph,q\varepsilon}^{-conf}(W) = f_{B,0T_c} \\ N_{ph,q\varepsilon}^{+conf}(0) &= N_{ph,q\varepsilon}^{-conf}(0) \end{aligned} \tag{4.10}$$

where $f_{B,0Tc}$ denotes the Bose-Einstein distribution at zero chemical potential and temperature T_c. The respective solutions to Eq. (4.10) are

$$\begin{aligned} N_{ph,q\varepsilon}^{+esc}(z) &= \left(N_{inc}(\theta_q, \varphi_q) + f_{B,0T_c}\right)\left(1 - \alpha_{q,\varepsilon}^{\max}(f_l - f_u)\frac{z}{|\cos\theta_q|}\right) + f_{B,\mu T_c}\alpha_{q,\varepsilon}^{\max}(f_l - f_u)\frac{z}{|\cos\theta_q|} \\ N_{ph,q\varepsilon}^{-esc}(z) &= f_{B,0T_c}\left(1 - \alpha_{q,\varepsilon}^{\max}(f_l - f_u)\frac{W-z}{|\cos\theta_q|}\right) + f_{B,\mu T_c}\alpha_{q,\varepsilon}^{\max}(f_l - f_u)\frac{W-z}{|\cos\theta_q|} \\ N_{ph,q\varepsilon}^{-conf}(z) &= f_{B,0T_c}\left(1 - \alpha_{q,\varepsilon}^{\max}(f_l - f_u)\frac{W-z}{|\cos\theta_q|}\right) + f_{B,\mu T_c}\alpha_{q,\varepsilon}^{\max}(f_l - f_u)\frac{W-z}{|\cos\theta_q|} \\ N_{ph,q\varepsilon}^{+conf}(z) &= f_{B,0T_c}\left(1 - \alpha_{q,\varepsilon}^{\max}(f_l - f_u)\frac{W+z}{|\cos\theta_q|}\right) + f_{B,\mu T_c}\alpha_{q,\varepsilon}^{\max}(f_l - f_u)\frac{W+z}{|\cos\theta_q|} \end{aligned} \tag{4.11}$$

keeping in mind that $\cos\theta_k < 0$ for modes travelling in the negative z direction. $\alpha_{q,\varepsilon}^{\max}(f_l - f_u) = \alpha_{q,\varepsilon}^{\max}[f_l(1 - f_u) - f_u(1 - f_l)]$ is now recognised as the net absorption coefficient for the given transition (such that $\alpha_{q,\varepsilon}^{\max}$ is the maximum possible absorption coefficient for this transition, achieved when the electron occupancy of the lower and upper level is 1 and 0 respectively). We can see from Eq. (4.11) that the photons enter the IB layer with the external photon population and, over the course of their trajectory, move toward equilibrium with the electron states (with chemical potential μ) at a rate determined by the absorption coefficient, as is discussed in, e.g., Ref. [13].

The density of polarized photon modes per unit volume per unit solid angle per unit energy interval inside the absorber is

$$\rho_{phot} = \frac{2n_{ref}^3 \left(E_{phot}\right)^2}{h^3 c^3} \tag{4.12}$$

The total net generation current density for the $l \rightarrow u$ transition is the difference between the total number of photons entering and escaping the cell per unit area in the xy plane. Taking into account photon fluxes at both interfaces, net generation current density per unit energy interval is

$$J_{l\rightarrow u} = e \int_0^{E_g} dE_{phot} \frac{c}{n_{ref}} \rho_{phot} \left[\begin{array}{l} \int_0^{2\pi} \int_0^{\theta_c} d\varphi d\theta |\cos\theta| \sin\theta \left\langle N_{ph,q\varepsilon}^{+esc}(0) - N_{ph,q\varepsilon}^{+esc}(W) \right\rangle_{\varepsilon} \\ - \int_0^{2\pi} \int_{\theta_c}^{\pi/2} d\varphi d\theta |\cos\theta| \sin\theta \left\langle N_{ph,q\varepsilon}^{+conf}(W) \right\rangle_{\varepsilon} \\ + \int_0^{2\pi} \int_{\pi/2}^{\pi-\theta_c} d\varphi d\theta |\cos\theta| \sin\theta \left\langle N_{ph,q\varepsilon}^{-conf}(W) \right\rangle_{\varepsilon} \\ + \int_0^{2\pi} \int_{\pi-\theta_c}^{\pi} d\varphi d\theta |\cos\theta| \sin\theta \left\langle N_{ph,q\varepsilon}^{-esc}(W) - N_{ph,q\varepsilon}^{-esc}(0) \right\rangle_{\varepsilon} \end{array} \right] \tag{4.13}$$

where $\langle\rangle_{\varepsilon}$ represents an averaging over polarization states. After some mathematical handling this leads to

$$J_{l\rightarrow u} = e \int_0^{E_g} dE_{phot} \frac{c}{n_{ref}} \rho_{phot} W(f_l - f_u) \left[\begin{array}{l} \int_0^{2\pi} \int_0^{\theta_c} d\varphi d\theta \sin\theta \left\langle \alpha_{q,\varepsilon}^{\max} \right\rangle_{\varepsilon} N_{inc} \\ + \left(f_{B,0T_c} - f_{B,\mu T_c}\right) \int_0^{2\pi} \int_0^{\pi} d\varphi d\theta \sin\theta \left\langle \alpha_{q,\varepsilon}^{\max} \right\rangle_{\varepsilon} \end{array} \right] \tag{4.14}$$

where it is assumed that the incident flux is unpolarized.

To compute (4.14), it is necessary to consider the polarization dependence of $\alpha_{q,\varepsilon}^{\max}$. Expanding the square modulus of the optical matrix element we obtain Eq. 2.73, which is repeated here,

$$\begin{aligned} |\langle \Xi_l | \mathbf{r} \cdot \boldsymbol{\varepsilon} | \Xi_u \rangle|^2 = {} & \varepsilon_x^2 |\langle \Xi_l | x | \Xi_u \rangle|^2 + \varepsilon_y^2 |\langle \Xi_l | y | \Xi_u \rangle|^2 + \varepsilon_z^2 |\langle \Xi_l | z | \Xi_u \rangle|^2 \\ & + 2\varepsilon_x \varepsilon_y |\langle \Xi_l | x | \Xi_u \rangle| |\langle \Xi_l | y | \Xi_u \rangle| + 2\varepsilon_y \varepsilon_z |\langle \Xi_l | y | \Xi_u \rangle| |\langle \Xi_l | z | \Xi_u \rangle| \\ & + 2\varepsilon_z \varepsilon_x |\langle \Xi_l | z | \Xi_u \rangle| |\langle \Xi_l | x | \Xi_u \rangle| \end{aligned} \tag{4.15}$$

We can write three separate absorption coefficients:

$$\begin{aligned}
\alpha_x^{\max} &= \frac{2\pi^2 e^2 \hbar\omega}{n_{ref} c h \varepsilon_0} \rho_{QD} |\langle \Xi_l | x | \Xi_u \rangle|^2 \delta(E_u - E_l - E_{phot}) \\
\alpha_y^{\max} &= \frac{2\pi^2 e^2 \hbar\omega}{n_{ref} c h \varepsilon_0} \rho_{QD} |\langle \Xi_l | y | \Xi_u \rangle|^2 \delta(E_u - E_l - E_{phot}) \\
\alpha_z^{\max} &= \frac{2\pi^2 e^2 \hbar\omega}{n_{ref} c h \varepsilon_0} \rho_{QD} |\langle \Xi_l | z | \Xi_u \rangle|^2 \delta(E_u - E_l - E_{phot})
\end{aligned} \tag{4.16}$$

such that

$$\begin{aligned}
\alpha_{q,\varepsilon}^{\max} &= \varepsilon_x^2 \alpha_x^{\max} + \varepsilon_y^2 \alpha_y^{\max} + \varepsilon_z^2 \alpha_z^{\max} \\
&+ 2\varepsilon_x \varepsilon_y \sqrt{\alpha_x^{\max} \alpha_y^{\max}} + 2\varepsilon_y \varepsilon_z \sqrt{\alpha_y^{\max} \alpha_z^{\max}} + 2\varepsilon_z \varepsilon_x \sqrt{\alpha_z^{\max} \alpha_x^{\max}}
\end{aligned} \tag{4.17}$$

We now wish to compute $\langle \alpha_{q,\varepsilon}^{\max} \rangle_\varepsilon$: the average over all polarization states for the photon mode in question. To do this, we define two mutually orthogonal unit vectors in the polarization plane of the photon mode. The simplest choice is the vectors $\boldsymbol{s}$ and $\boldsymbol{p}$ where

$$\begin{aligned}
\boldsymbol{s} &= \sin\varphi_k \hat{\boldsymbol{x}} - \cos\varphi_k \hat{\boldsymbol{y}} \\
\boldsymbol{p} &= \cos\varphi_k \cos\theta_k \hat{\boldsymbol{x}} + \sin\varphi_k \cos\theta_k \hat{\boldsymbol{y}} - \sin\theta_k \hat{\boldsymbol{z}}
\end{aligned} \tag{4.18}$$

where $\hat{\boldsymbol{x}}$, $\hat{\boldsymbol{y}}$, $\hat{\boldsymbol{z}}$ are unit vectors in the x, y and z directions, and θ_k, φ_k are the polar and azimuth angles of the mode's wavevector (not of the polarization vector). We can now calculate $\langle \alpha_{q,\varepsilon}^{\max} \rangle_\varepsilon$ by taking the average of $\alpha_{q,\varepsilon}^{\max}$ for all polarizations included in the $\boldsymbol{sp}$ plane. The generic polarization vector is $\boldsymbol{\varepsilon} = \boldsymbol{s}\cos\gamma + \boldsymbol{p}\sin\gamma$, γ being an arbitrary angle. This generic polarization vector is written as a linear function of $\hat{\boldsymbol{x}}, \hat{\boldsymbol{y}}, \hat{\boldsymbol{z}}$ according to Eq. (4.18) and then its projection factors $\varepsilon_x^2, \varepsilon_y^2, \varepsilon_z^2, \varepsilon_x\varepsilon_y, \varepsilon_y\varepsilon_z, \varepsilon_z\varepsilon_x$ appearing in Eq. (4.17) are averaged by integrating each factor over all the possible $0 \leq \gamma < 2\pi$ and dividing by 2π. The result is,

$$\begin{aligned}
\left\langle \alpha_{\mathbf{k},\varepsilon}^{\max} \right\rangle_\varepsilon &= \frac{1 - \cos^2\varphi_k \sin^2\theta_k}{2} \alpha_x^{\max} + \frac{1 - \sin^2\varphi_k \sin^2\theta_k}{2} \alpha_y^{\max} + \frac{\sin^2\theta_k}{2} \alpha_z^{\max} \\
&- \cos\varphi_k \sin\varphi_k \sin^2\theta_k \sqrt{\alpha_x^{\max} \alpha_y^{\max}} - \cos\varphi_k \cos\theta_k \sin\theta_k \sqrt{\alpha_y^{\max} \alpha_z^{\max}} \\
&- \sin\varphi_k \cos\theta_k \sin\theta_k \sqrt{\alpha_z^{\max} \alpha_x^{\max}}
\end{aligned} \tag{4.19}$$

We can now compute Eq. (4.14). We first observe that

$$\int_0^{2\pi}\int_0^{\pi} d\varphi d\theta \sin\theta \frac{1-\cos^2\varphi\sin^2\theta}{2} = \int_0^{2\pi}\int_0^{\pi} d\varphi d\theta \sin\theta \frac{1-\sin^2\varphi\sin^2\theta}{2}$$
$$= \int_0^{2\pi}\int_0^{\pi} d\varphi \frac{\sin^3\theta}{2} = \frac{2}{3} \tag{4.20}$$

and that

$$\frac{c}{n_{ref}}\rho_{phot}\int_0^{2\pi}\int_0^{\theta_c} d\theta \sin\theta \frac{1-\cos^2\varphi\sin^2\theta}{2} N_{inc} \approx \frac{c}{n_{ref}}\rho_{phot}\frac{1}{2}\int_0^{2\pi}\int_0^{\theta_c} d\theta \sin\theta\cos\theta N_{inc} = \frac{F_{ph,inc}}{2}$$
$$\frac{c}{n_{ref}}\rho_{phot}\int_0^{2\pi}\int_0^{\theta_c} d\theta \sin\theta \frac{1-\sin^2\varphi\sin^2\theta}{2} N_{inc} \approx \frac{F_{ph,inc}}{2}$$
$$\frac{c}{n_{ref}}\rho_{phot}\int_0^{2\pi}\int_0^{\theta_c} d\theta \frac{\sin^3\theta}{2} N_{inc} = 0 \tag{4.21}$$

where $F_{ph,inc}$ is the incident photon flux per unit time and area and we have made the approximations $\sin^2\theta \approx 0$ and $1 \approx \cos\theta$, which are valid for $\theta < \theta_c$. Finally, we observe that

$$\int_0^{2\pi}\int_0^{\pi} d\varphi d\theta \cos\varphi \sin\varphi \sin^3\theta = \int_0^{2\pi}\int_0^{\theta_c} d\varphi d\theta \cos\varphi \sin\varphi \sin^3\theta$$
$$= \int_0^{2\pi}\int_0^{\pi} d\varphi d\theta \cos\varphi \cos\theta \sin^2\theta = \int_0^{2\pi}\int_0^{\theta_c} d\varphi d\theta \cos\varphi \cos\theta \sin^2\theta$$
$$= \int_0^{2\pi}\int_0^{\pi} d\varphi d\theta \sin\varphi \cos\theta \sin^2\theta = \int_0^{2\pi}\int_0^{\theta_c} d\varphi d\theta \sin\varphi \cos\theta \sin^2\theta = 0 \tag{4.22}$$

Equation (4.14) is therefore

$$J_{l\to u} = eW\int_0^{E_g} dE_{phot}(f_l - f_u)\left(\begin{array}{l}\frac{F_{ph,inc}}{2}\left(\alpha_x^{\max}+\alpha_y^{\max}\right)\\ +\frac{8\pi n_{ref}^2 E_{phot}^2}{3h^3c^2}\left(f_{B,0T_c}-f_{B,\mu T_c}\right)\left(\alpha_x^{\max}+\alpha_y^{\max}+\alpha_z^{\max}\right)\end{array}\right)$$
$$= eW\int_0^{E_g} dE_{phot}(f_l - f_u)\left(F_{ph,inc}\alpha_{x,y}^{\max} + \frac{8\pi n_{ref}^2 E_{phot}^2}{h^3c^2}\left(f_{B,0T_c}-f_{B,\mu T_c}\right)\alpha_{iso}^{\max}\right) \tag{4.23}$$

where $\alpha_{x,y}^{max} = \left(\alpha_x^{max} + \alpha_y^{max}\right)/2$ is the average absorption coefficient for photons polarized in the xy plane, and $\alpha_{iso}^{max} = \left(\alpha_x^{max} + \alpha_y^{max} + \alpha_z^{max}\right)/3$ is the average absorption coefficient for isotropic photons.

Equation (4.23) contains a term that is dependent on the incident photon flux and a term that is not. It is therefore convenient to decompose $J_{l\to u}$ into an illumination current, $J_{l\to u}^L$, and a dark current, $J_{l\to u}^D$.

$$\begin{aligned} J_{l\to u} &= J_{l\to u}^L + J_{l\to u}^D \\ J_{l\to u}^L &= eW\int_0^{E_g} dE_{phot}(f_l - f_u)F_{ph,inc}\alpha_{x,y}^{max} \\ J_{l\to u}^D &= eW\int_0^{E_g} dE_{phot}(f_l - f_u)\frac{8\pi n_{ref}^2 E_{phot}^2}{3h^3c^2}\left(f_{B,0T_c} - f_{B,\mu T_c}\right)\alpha_{iso}^{max} \end{aligned} \tag{4.24}$$

In the expression for the dark current, the term proportional to $f_{B,0Tc}$ represents upward transitions due to the absorption of thermal photons incident on the IB layer from the surroundings; this is later described as radiative thermal escape. The term proportional to $-f_{B,\mu Tc}$ represents downward transitions due to the emission of luminescent photons into the surroundings. Since $\mu = E_{F,u} - E_{F,l}$, we can see that the dark current between two levels will always be a net flow from the level with higher QFL to the level with lower QFL, which is thermodynamically consistent.

Equation (4.24) gives the net generation current between a pair of IB levels as a function of the quantities α_x^{max} etc. which are directly calculated from the optical matrix elements via Eq. (4.16). The optical matrix elements were calculated in Chaps. 2 and 3.

The net current from an IB level to the CB can be calculated using Eq. (4.24) and modifying the absorption coefficient of Eq. (4.16) as follows:

$$\alpha_x^{max}(E_{phot}) = \frac{2\pi^2 e^2 E_{phot}}{n_{ref} c h \varepsilon_0}\rho_{QD}\sum_{k\in CB}|\langle\Xi_l|x|\Xi_k\rangle|^2\delta(E_k - E_l - E_{phot}) \tag{4.25}$$

where the sum is made over all possible final states in the CB. Similar expressions can be written for transitions from the VB to an IB level, or from the VB to the CB.

Finally, the reader is reminded that $J_{l\to u}^L$ in Eq. (4.24) has been derived for unpolarised illumination, such as sunlight. For linearly polarized illumination, where the polarization makes an angle γ with the x direction, the illumination current is obtained by replacing $\alpha_{x,y}^{max}$ with $\alpha_x^{max}\cos\gamma + \alpha_x^{max}\sin\gamma$ in the expression for $J_{l\to u}^L$ in Eq. (4.24). For example, a polarized laser source can be treated my making this change and taking $F_{ph,inc}$ to be the spectral distribution of the laser line. No change is required in the expression for $J_{l\to u}^D$, since it is independent of the incident illumination.

4.2.2 Charge Neutrality Condition

Charge neutrality is assumed to in the QD stack. This condition is found in the inside (far from the edges) of thick stacks of QDs, as is the case for the QD-IBSC under investigation [14]. We can therefore write the equation

$$n_{CB} - p_{VB} + n_{IB} - N_d^+ = 0 \tag{4.26}$$

where n_{CB} and p_{VB} are the concentrations of electrons and holes in the CB and VB respectively (taking also the confined hole states into account), n_{IB} is the electron concentration in the several IB states (summed), N_d^+ is the concentration of ionized donor atoms used to prefill the IB and E_d is the donor level energy.

The relevant quantities are given by

$$\begin{aligned}
n_{CB} &= N_{CB}\mathrm{e}^{-(E_{CB}-E_{F,CB})/k_BT_c} \\
p_{VB} &= N_{VB}\mathrm{e}^{-(E_{F,VB}-E_{VB})/k_BT_c} + \sum_{\text{confined hole states}} \frac{2\rho_{QD}}{\mathrm{e}^{(E_{F,VB}-E_{\text{hole state}})/k_BT_c}+1} \\
n_{IB} &= \sum_{IB} \frac{2\rho_{QD}}{\mathrm{e}^{(E_{IB}-E_{F,IB})/k_BT_c}+1} \\
N_d^+ &= \frac{N_d}{\mathrm{e}^{(E_{F,CB}-E_d)/k_BT_c}+1}
\end{aligned} \tag{4.27}$$

where N_{CB} and N_{VB} are the effective densities of states in the CB and VB respectively. These are

$$\begin{aligned}
N_{CB} &= 2\left(\frac{2\pi m_e^* k_B T_C}{h^2}\right)^{\frac{3}{2}} \\
N_{VB} &= 2\left(\frac{2\pi m_h^* k_B T_C}{h^2}\right)^{\frac{3}{2}}
\end{aligned} \tag{4.28}$$

where m_e^* and m_h^* are the effective masses for electrons and holes, which are taken in our exemplary cell to be those for GaAs at the Γ point.

In the second and third lines of Eq. (4.27), the sums are respectively made over all confined hole states in the VB pedestal and confined electron states in the CB well (the latter are the IB levels). The factor of 2 accounts for spin degeneracy. The summand must also be multiplied to account for *xy* degeneracies where appropriate, for example for the IB(121)/IB(211) level.

Finally N_d is the concentration of donor atoms used to prefill the IB and E_d is the donor energy level. The QD layer of the exemplary QD-IBSC is δ-doped with Si to a concentration that supplies one donor atom per QD [3]. We therefore take $N_d = \rho_{QD}$ and $E_d = E_{CB} - 0.006$ eV [15]. In all our modelling results, the donors have been found to be completely ionized, even at low temperatures. This is because, at zero bias (as is appropriate for quantum efficiency experiments), $E_{F,CB}$ is bound to the IB levels, which are below the donor level.

4.2.3 The Terminal Voltage

Finally, we make the simplifying assumption that the terminal voltage is equal to the split between the VB and CB QFLs. This is equivalent to assuming infinite carrier mobility and, therefore, zero series resistance.

$$E_{F,CB} - E_{F,VB} = q_e V \tag{4.29}$$

4.2.4 Solution of the Problem to Find the Output Current

Through Eqs. (4.1), (4.28) and (4.29), we have a system of simultaneous equations with as many equations as unknowns. The input parameters are the energy levels of both confined and extended states in all bands, the Fermi-level-free absorption coefficients ($\alpha^{\max}$) for all possible transitions, the cell temperature, the terminal voltage and the illumination photon flux. The problem is solved numerically using Wolfram Mathematica®, yielding the QFLs. The net generation current for each transition is then calculated using Eq. (4.23).

In order to calculate the total current extracted from the SOTA cell, we must sum the contributions from all transitions and chains of transitions that begin in the host VB and end in the host CB. These can be divided into three contributing currents, as is shown in Eq. (4.30):

$$J_{total} = J_{IB} + J_{VBpedestal\text{–}\mathrm{CB}} + J_{above\text{–}bandgap} \tag{4.30}$$

J_{IB} is the net current from the VB to the CB via the IB levels. This is the sum of the transition currents from the VB to the IB levels, or, equivalently, the sum of the transition currents from the IB levels to the CB.

$$\begin{aligned} J_{IB} &= J_{VB\text{–}\mathrm{IB}(111)} + J_{VB\text{–}\mathrm{IB}(211)} + J_{VB\text{–}\mathrm{IB}(221)} \\ &= J_{IB(111)\text{–}\mathrm{CB}} + J_{IB(211)\text{–}\mathrm{CB}} + J_{IB(221)\text{–}\mathrm{CB}} \end{aligned} \tag{4.31}$$

where each transition current ($J_{VB\text{-}IB(111)}$ etc.) is calculated using Eq. (4.23).

$J_{VBpedestal\text{-}CB}$ is the current that flows directly from the confined hole states in the VB pedestal to the states in the energy range of the host CB. The reader is reminded that holes generated in the VB pedestal are assumed to undergo quick thermal escape to the host VB and are therefore counted as extracted current. $J_{VBpedestal\text{-}CB}$ is a subbandgap current and is calculated directly from Eq. (4.23) using the absorption coefficient shown as a black curve in Fig. 4.2.

Finally, $J_{above\text{-}bandgap}$ is the net current generated by direct transitions from the host VB to the host CB. The QD stack is sandwiched between layers of the host material (GaAs) that are thick enough to absorb all above-bandgap photons. We can therefore assume $J_{above\text{-}bandgap}$ to be equal to the current generated by a single-gap

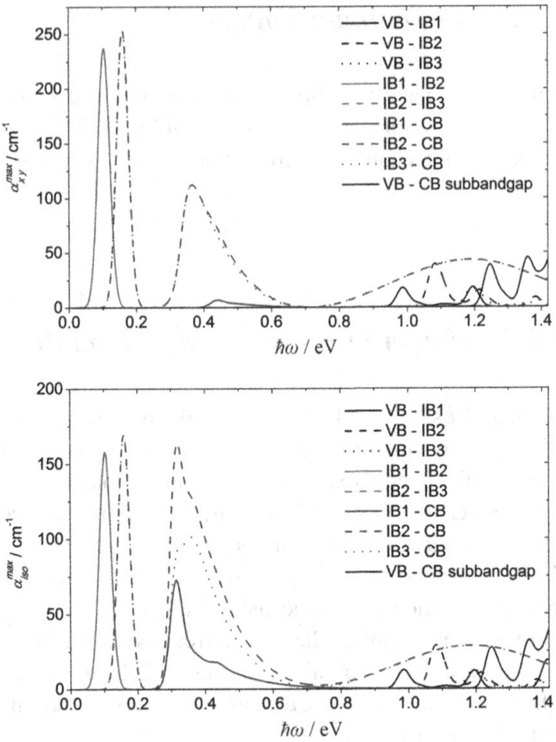

Fig. 4.2 Absorption coefficients for the different transitions in the exemplary QD-IBSC before modification by the electron occupancies of the lower and upper levels. *Upper panel* absorption coefficients for photons polarized in the xy plane. *Lower panel* absorption coefficient for photons polarized isotropically in the three dimensions. Each *curve* represents a different electronic transition between all the bands and levels shown in Fig. 4.1. Reproduced with permission. © 2014, Wiley [10]

GaAs solar cell, and calculate it using the standard Shockley-Queisser detailed balance model for single-gap solar cells [2, 16]. $J_{above\text{-}bandgap}$ can also be thought of as the current that would be generated by the solar cell if the QDs were not present.

4.3 Input Parameters

For the SOTA cell, the optical matrix elements have been calculated in the previous chapters and in Refs. [5–9]. The resulting $\alpha_{x,y}^{\max}$ and $\alpha_{iso}^{\max}$ are calculated from Eq. (4.16), remembering that $\alpha_{x,y}^{\max} = \left(\alpha_x^{\max} + \alpha_y^{\max}\right)/2$ and $\alpha_{iso}^{\max} = \left(\alpha_x^{\max} + \alpha_y^{\max} + \alpha_z^{\max}\right)/3$, and are used as the input for the calculations; these are plotted in Fig. 4.2. The Dirac-delta in Eq. (4.16) has been approximated as a Gaussian with a standard deviation of 25 meV to account for the variation of QD sizes in the layer stack. The IB(111) $\leftrightarrow$ IB (221) transition is forbidden for all polarizations [5]; hence, this is not plotted. "VB-CB sub-bangap" refers to transitions between the pseudo continuum of states in the VB pedestal and the CB, the transition energies for these being lower than the host bandgap E_g. Above the bandgap the absorption is several orders of magnitude higher. The remaining input parameters used in the detailed balance model are listed in Table 4.1.

Table 4.1 Input parameters used in detailed balance model

Symbol	Value	Units
E_{CB}	0	eV
E_{VB} at 300 K	−1.42	eV
$E_{VB,eff}$ at 300 K	−1.24	eV
$E_{IB(111)}$	−0.26	eV
$E_{IB(211)}$	−0.16	eV
$E_{IB(221)}$	−0.06	eV
W	2.4	μm
n_{ref}	3.5	
ρ_{QD}	5×10^{15}	cm^{-3}
N_d	5×10^{15}	cm^{-3}
E_d	−0.006	eV
N_{CB}	$8.4 \times 10^{13}\ T_c^{3/2}$	cm^{-3}
N_{VB}	$1.8 \times 10^{15}\ T_c^{3/2}$	cm^{-3}

The energy origin is at the host CB band edge. T_c is expressed in degrees Kelvin

4.4 The Internal Quantum Efficiency of the SOTA Cell

4.4.1 Analysis of the Internal Quantum Efficiency

The model has been used to simulate the experiments presented in Refs. [3, 4], in which the internal quantum efficiency (IQE) of the SOTA cell was measured at different temperatures. Before presenting the modelled results, we comment briefly on the relevance of the experimental results. In the articles, it was demonstrated that electrons pumped from the VB to the IB by external illumination can readily escape at room temperature to the CB by thermal processes. Strong IB ↔ CB thermal escape is thought to imply that positive sub-bandgap photocurrent can only be delivered at a voltage limited by the VB-IB sub-bandgap [17], which is smaller than the overall VB-CB bandgap. This ultimately limits the conversion efficiency of the device. A goal of QD-IBSC research has since been to minimise the thermal escape, for example by using a larger bandgap hosts [18, 19].

In the experiment, the temperature of the QD-IBSC and its immediate surroundings are controlled using a cryostat and the QD-IBSC was illuminated by a monochromator with a linewidth of around 1 nm and a total irradiance of 0.5 mW cm^{-2}. In the model, the temperature is controlled by the parameter T_c, and the incident flux is modelled as a narrow Gaussian function with a deviation of $\sigma = 0.01$ eV.

$$\Phi_{inc}(E_{phot}) = \frac{\exp\left(-\left(E_{phot} - E_{mon}\right)^2 / \sigma\right)}{\sigma\sqrt{\pi}} \times 0.5\ \text{mW cm}^{-2} \tag{4.32}$$

where E_{mon} is the nominal output photon energy of the monochromator.

To calculate the IQE, the model described in Sect. 4.2 is applied for a range of monochromator photon energies, E_{mon} . Zero bias voltage is assumed. The calculated total current, J_{total}, is then normalised by the total incident photon flux (times e) to yield the dimensionless IQE.

Figure 4.3 (left) shows the calculated IQE for a range of cell temperatures. Figure 4.3 (right) shows the measured temperature dependent IQE published in Refs. [3, 4]. Note that, in both graphs, the 'Photon Energy' on the horizontal axis refers to the nominal output photon energy of the monochromator. The peaks marked in the experimental curve as E_0 and E_1 are transitions from the VB to the ground and first excited states in the QD CB well; these correspond to states IB (1,1,1) and IB(2,1,1)/(1,2,1) in the model. The quantitative agreement between the experimental IQE and that calculated using the EKPH method for energies below 1.35 eV and at room temperature has been discussed in Refs. [7, 8, 10]. The fact that the weak sub-bandgap IQE has been reproduced by the model, which assumes infinite carrier mobility, demonstrates that this is a problem of weak photon absorption in the QDs and not of carrier extraction from the QD stack. The detailed balance model also reproduces qualitatively the temperature dependence of the IQE

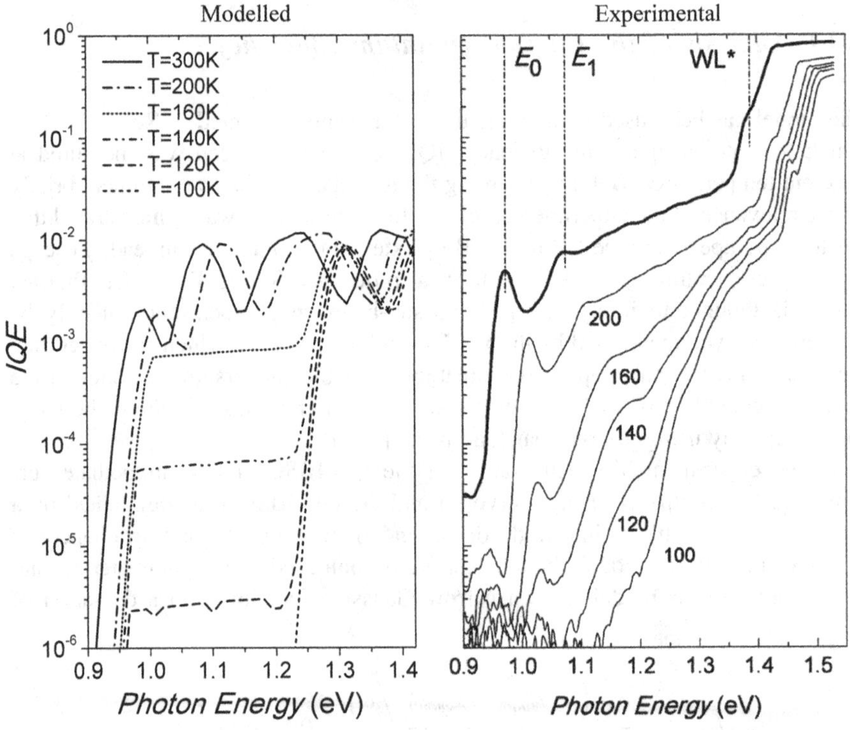

Fig. 4.3 Temperature dependent IQE for the exemplary QD-IBSC. *Left* IQE calculated using the detailed balance model. *Right* Measured IQE from Refs. [3, 4]. The photon energy on the *horizontal scale* refers to the nominal output photon energy of the monochromator [E_{mon} in Eq. (4.32)] Reproduced with permission. © 2014, Wiley [10]

in the 0.9–1.2 eV range. The quantitative agreement of this dependence is discussed in the Sect. 4.4.2.

The modeled IQE ceases to be temperature dependent above around 1.2 eV. This is the effective bandgap of the IB layer caused by the quasi continuum of confined-hole states that invade the bandgap (see Fig. 3.3). This can be seen somewhat in the measured data at lower temperatures. Both modeled and experimental data show a blue shifting of features at lower temperatures, caused by the widening of the host bandgap.

There are also features not reproduced by the model. The step in the experimental data around 1.35 eV (labelled WL) is due to absorption by the quantum wells formed by the so-called wetting layer during QD growth. Also, the measured IQE increases steadily with increasing photon energy, an effect that becomes more pronounced at lower temperatures. This could be due to non-instantaneous relaxation of holes from the host valence band to the confined hole states in the QDs or to the non-applicability of the model to states with high quantum numbers due to inaccuracy of the parabolic band model and to failure of the integral factorization rule. Neither of these effects has been considered in the model.

4.4.2 Arrhenius Plot

To provide quantitative analysis, an Arrhenius plot of the modelled IQE at the E_0 peak is shown in Fig. 4.4 (left) (squares), along with the measured Arrhenius plot from Ref. [3] in Fig. 4.4 (right). Both the modelled and experimental data show distinct high and low temperature regimes.

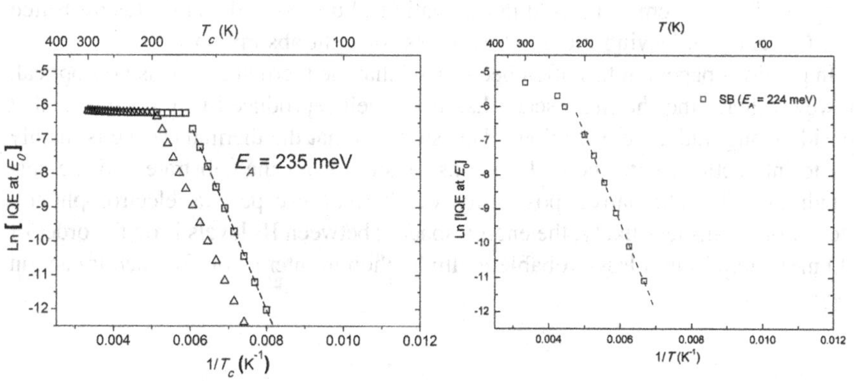

Fig. 4.4 Arrhenius plots of the IQE at E_0 for the exemplary QD-IBSC. *Left graph* values calculated using detailed balance model. *Squares* monochromator irradiance = 0.5 mW cm^{-2}. *Triangles* monochromator irradiance = 5 mW cm^{-2}. *Right* measured values published in Ref. [3]. In both graphs, the *dashed lines* are linear fits to the linear parts of the curves; the thermal activation energies E_A are extracted from the slopes of these fits. Reproduced with permission. © 2014, Wiley [10]

In the high temperature regime, the photocurrent is limited by VB → IB transitions induced by the monochromator illumination. The modelled data is entirely temperature independent in this regime, whereas the experimental appears to approach a plateau gradually on increasing the temperature. This discrepancy is also visible in Fig. 4.3 (comparing results at 300 and 200 K). We consider there is a basic agreement; small discrepancies are to be expected given the simplicity of the model.

In both plots, the low temperature regime is linearly temperature dependent. Here the limitation is due to thermal escape from the IB to the CB. The linearity implies an Arrhenius type temperature dependence of the form $\exp(-E_A/k_BT_c)$, where E_A is the so-called activation energy of the thermal escape. Applying a linear fit to the modelled data in this regime yields and activation energy of 235 meV, to be compared to the 224 eV yielded from the experimental data [3]. In the model, the thermal escape occurs due to absorption of the aforementioned thermal photons incident on the IB layer from the surroundings and from the substrate, as was postulated in Ref. [20]. Due to the good agreement between the measured thermal escape activation energy and that predicted by the model, we believe the model can serve as a useful tool for evaluating future proposals for QD geometries and material systems for QD-IBSCs.

The calculation has been repeated with a monochromator irradiance of 5 mW cm^{-2}, ten times higher than the estimated irradiance in the experiment. The results are plotted as triangles in Fig. 4.4 (left). The change in irradiance causes a lateral shift in the temperature dependent part of the curve, achieving much better agreement with the experimental data. Although it is possible that the experimental irradiance is actually higher than estimated, this discrepancy is more likely due to the calculated E_0 absorption peak being below the experimental one (as is visible in Fig. 4.3). This may be due the prefilling of the IB fundamental state by doping being not 5×10^{15} cm^{-3}, as used in calculations, but 1.8×10^{15} cm^{-3} (as measured with CV curves), leaving more empty states for light absorption.

In previous papers, it has often been stated that the thermal escape is non-optical. However, here, the thermal escape has been well reproduced using a model that considers only radiative transitions. This suggests that the thermal escape is mainly due to interactions with thermal photons incident from the substrate and the surroundings. The alternative possibility of thermal escape via electron-phonon interaction seems less likely; the energy spacing between IB levels is on the order of 100 meV, requiring a less-probable multiple phonon interaction for each transition to occur.

4.4.3 Analysis of the Individual Transitions

Using the model, we can investigate the individual sub-bandgap currents that lead to the overall IQE. We define the IQE for each sub-bandgap transition as being the net number of charge carriers making that transition per incident photon, where positive and negative IQEs corresponds to net generation and recombination respectively. The IQE for each individual transition is plotted in Fig. 4.5 (left) and (right) for temperatures of 300 and 140 K respectively, these belonging to the two different regimes. Each figure is divided into three panels: the top panel shows transitions from the VB to the IB levels, the middle panel transitions between IB levels, and the bottom panel transitions from IB levels to the CB. Table 4.2 summarises the dominant path by which electrons are delivered from the VB to the CB in each energy range, these turn out to be the same for both temperatures.

At 300 K, the monochromator pumps electrons from the VB to one of the IB levels depending on the photon energy. The electrons in this level then make a transition to the IB(211) level due to absorption/emission of thermal photons. From the IB(211) level they are extracted to the CB, again due to absorption of thermal photons.

At 140 K, the dominant paths by which electrons reach the CB via the IB states is the same as for 300 K. However, due to the reduced radiative thermal escape at this temperature, the IB(211) → CB current is much weaker than at 300 K (note the different vertical scales). Consequently, most of the electrons reaching the IB levels from the VB recombine back to the VB via the IB(111) level.

These results reveal why the measurement of optical transitions from the IB ground state to the CB has as yet been very elusive. For normally incident photons

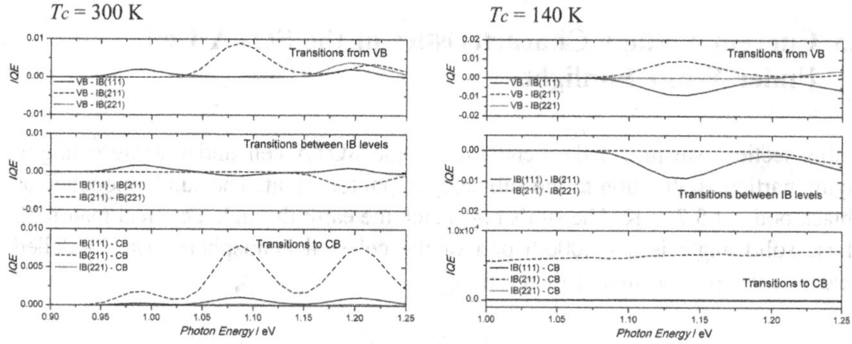

Fig. 4.5 IQEs of the individual transitions in the multi-level system for the exemplary QD-IBSC cell at different temperatures. *Left* T_c = 300 K. *Right* T_c = 140 K. The IQE here means the net number of charge carriers making the stated transition per incident photon. Negative values imply net recombination for the transition. Reproduced with permission. © 2014, Wiley [10]

Table 4.2 Dominant paths by which electrons reach the CB via the IB states for different monochromator energy ranges. The paths are the same at 300 K and 140 K

Energy range (eV)	Dominant path
0.95–1.02	VB → IB(111) → IB(211) → CB
1.02–1.15	VB → IB(211) → CB
1.15–1.35	VB → IB(221) → IB(211) → CB

(x, y polarization), there is weak direct coupling from the IB(111) level to the CB (Fig. 4.2, left). Hence, photogeneration via the IB(111) level is not a two-photon process but a three-photon process. Measuring this process at low temperatures would require photons of appropriate energy for each of the three transitions listed in the first row of Table 4.2. What has probably been often lacking in previous experiments is illumination by low energy photons (around 100 meV) to pump from the IB(111) to the IB(211) level.

Since radiative thermal escape always occurs via the IB(211) level and never directly from the ground IB(111) level, it seem reasonable that the excited IB(211) and IB(221) states must be removed from the forbidden band to suppress the escape. One suggested means of doing this is to decrease the QD size, as discussed in the previous chapters and in Refs. [21, 22]. This increases the energy of all IB levels, pushing the IB(211) and IB(221) states into the host CB, where they become virtual bound states [23]. However, the disadvantage is that the IB(111) energy also increases. This would place the IB(111) $\leftrightarrow$ CB transition at an energy at which thermal photons outnumber photons incident from the sun, and hence induce direct thermal escape from the IB(111) level. To effectively suppress thermal escape, it is therefore necessary to move to a material system with a larger CB offset, such as InAs/AlGaAs [18] or certain IV-VI/II-VI combinations [19], whilst simultaneously tuning the QD size so that all excited states are removed from the forbidden band.

4.5 Current Voltage Characteristics of the SOTA Cell Under White Sunlight

In this section, we model the behaviour of the SOTA cell under white sunlight, paying particular attention to the subbandgap photocurrent. The sun is modelled as a black body at 5,762 K. The sun's rays reach the earth through a conical manifold, whose solid angle is a 46,000th part of the celestial hemisphere. The modelled incident solar photon flux is therefore

$$F_{ph,inc}(\hbar\omega) = C\frac{1}{46{,}000}\frac{2\pi(\hbar\omega)^2}{h^3c^2}f_{B,0Ts}(\hbar\omega)\frac{1{,}000}{1{,}360} \tag{4.33}$$

where $f_{B,0Ts}(\hbar\omega)$ is the Bose-Einstein distribution of photons emitted from a body at the sun's photosphere temperature, Ts = 5,762 K. The factor of 1,000/1,360 is a correction to allow for absorption by the earth's atmosphere; this normalises the spectrum to a total irradiance of 1,000 Wm^{-2} as the usual standard at the Earth's surface (in the absence of concentration). C is the concentration factor, which accounts for any concentrating optics focusing sunlight on the cell and enlarging the cone of incidence. Throughout this chapter, we will take C = 1000X unless otherwise stated.

As in the previous section, the model described in Sect. 4.2 is used to determine the quasi Fermi levels (QFLs) of the three intermediate levels, as well as the VB and the CB, under a given set of conditions. These are then used to calculate the individual subbandgap transition currents via Eq. (4.23), and the total current generated by the SOTA cell is calculated via Eq. (4.30).

In the previous section, the terminal voltage V was set to zero in order to simulate a quantum efficiency experiment. Here, V is a changeable input parameter, allowing the current-voltage characteristics of the SOTA cell under white sunlight to be modelled. The reader is reminded that the model applies realistic absorption parameters, but is idealised in that it assumes the radiative limit. This was a realistic assumption for calculating the IQE in the previous section, since the SOTA cell is expected to work close to the radiative limit at short circuit. At higher bias voltages, it is believed that non-radiative mechanisms, particularly Shockley-Read-Hall recombination, become dominant in present QD-IBSC prototypes [15]. We would therefore expect the model to overestimate open circuit voltage (V_{oc}), and consequently overestimate efficiencies compared to present devices. However, these radiative-limit studies still serve to show fundamental mechanisms at higher bias voltages. Most non-radiative recombination mechanisms can be minimised by improving material quality; radiative recombination cannot.

4.5.1 JV Curve of the SOTA Cell: Individual Subbandgap Currents

As a first example, we consider the SOTA cell illuminated at 1000X concentration at a cell temperature of 300 K. The results are shown in Figs. 4.6, 4.7 and 4.8. Figure 4.6 shows the calculated total current, J_{total}, as well as the three constituent currents described in Eq. (4.30), as a function of the terminal voltage. At short circuit, the subbandgap currents (J_{IB} and $J_{VBpedestal\text{-}CB}$) are of much lower magnitude than $J_{above\text{-}bandgap}$ meaning that the presence of the QDs contributes only 0.12 A cm^{-2} to the short circuit current, J_{sc}, of the SOTA cell. The main effect of the QDs is therefore to lower the open circuit voltage, V_{oc}.

J_{IB} falls to zero at a lower voltage than $J_{VB\text{-}pedestal\text{-}CB}$ does. This indicates that the voltage at which positive current can be delivered via the IB is limited not by the effective bandgap $E_{g,eff}$, but by one of the smaller subbandgaps ($E_{IB(111)} - E_{VB,eff}$, $E_{IB(211)} - E_{VB,eff}$ or $E_{IB(221)} - E_{VB,eff}$). This is consistent with the strong radiative thermal escape from the IB levels to the CB that was predicted at short circuit in Sect. 4.4.

Due to the exponential nature of the dark current, one might expect the J_{total} to fall to zero almost immediately after J_{IB} does. However, from Fig. 4.6, the V_{oc} of the SOTA cell (the voltage at which J_{total} is zero) is predicted to be almost 300 meV higher than the voltage at which the J_{IB} falls to zero. The model therefore predicts that, in the radiative limit, the V_{oc} is limited neither by the subbandgaps nor by $E_{g,eff}$, but by some higher value. This is only possible because the subbandgap photon absorption is so weak. In the radiative model, only radiative recombination is taken into account. Both the rate of photon absorption and emission are proportional to the optical matrix

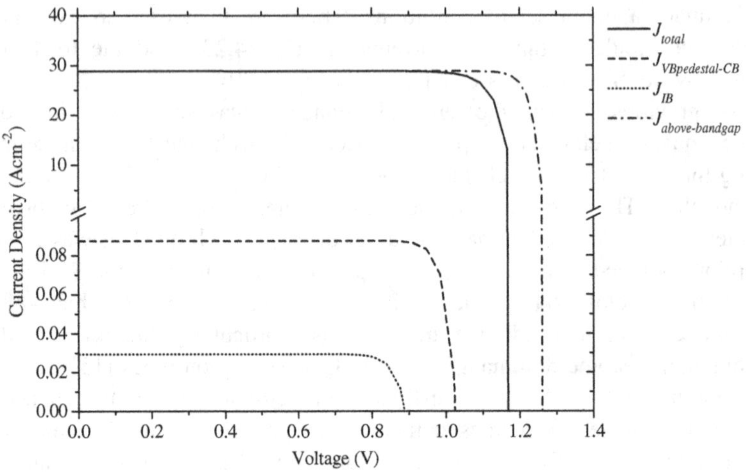

Fig. 4.6 The total current generated by the SOTA cell, J_{total}, as well as the three constituent currents described in Eq. (4.30). Note the *break* in the *vertical axis*, and the different scales before and after the break. The calculations were made using the detailed balance model for the SOTA cell at 1000X concentration and a cell temperature of 300 K

element (see Eq. 4.4). Hence, a weak subbandgap illumination current at short circuit implies a weak subbandgap dark current at higher voltages. Therefore the effect of the subbandgap dark current on the total current is small, and does not strongly limit the overall V_{oc}. Of course, for the SOTA cell to deliver high efficiency, it would need to have much stronger subbandgap photon absorption, in which case we would expect the V_{oc} to be more strongly limited. This is studied in Sect. 4.6.

The reader is reminded that all experimental samples to date are limited by non-radiative recombination close to open circuit; hence, their V_{oc} is much lower than that shown in Fig. 4.6. To our knowledge, the highest V_{oc} reported (to 2013) for an InAs/GaAs QD-IBSC in the literature is 0.997 V [24].

Figure 4.7 shows all the subbandgap transition currents via the IB levels as a function of the bias voltage. J_{IB} is also shown for reference. At short circuit, the VB $\rightarrow$ IB(211) $\rightarrow$ CB transition accounts for 80 % of the current generated via the IB levels. Interestingly, the model predicts that many subbandgap transitions exhibit net recombination (negative transition current), even at short circuit. It can be seen from the graph that many electrons pumped from the VB to the IB levels are returned to the VB by circular paths; the most dominant being VB $\rightarrow$ IB (211) $\rightarrow$ IB(111) $\rightarrow$ VB and VB $\rightarrow$ IB(221) $\rightarrow$ IB(211) $\rightarrow$ IB(111) $\rightarrow$ VB.

Figure 4.8 shows the quasi Fermi levels of the three IB levels and the two bands as a function of the bias voltage. It can be seen that the QFLs of the three IB levels are more or less pinned to the energy of the IB(111) level at all bias voltages. The ground state is therefore more or less half filled, and the excited levels have lower occupancies, although they are not completely empty, as evidenced by the non-zero $J_{211\text{-}CB}$ and $J_{221\text{-}CB}$ currents in Fig. 4.7.

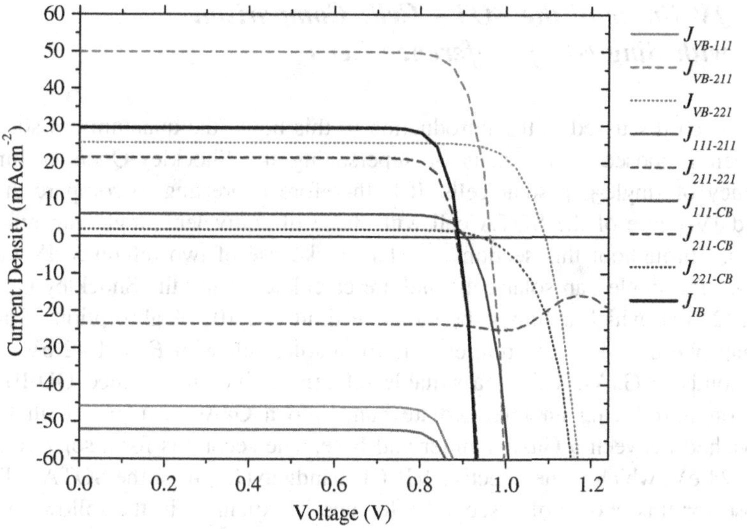

Fig. 4.7 Transition currents for the individual transitions via the IB levels as a function of the terminal voltage. The *curves* are colour coded; *red* for transitions from the VB to the IB levels, *green* for transitions between the IB levels and *blue* for transitions from the IB levels to the CB (increasing the *grey tone* in the printed text). The IB(1,1,1)–IB(2,2,1) transition is forbidden and is not plotted. Also shown is the total current generated via the IB, J_{IB}, where $J_{IB} = J_{VB\text{-}111} + J_{VB\text{-}211} + J_{VB\text{-}221} = J_{111\text{-}CB} + J_{211\text{-}CB} + J_{221\text{-}CB}$. The *horizontal line* represents zero net current. The calculations were made using the detailed balance model for the SOTA cell at 1000X concentration and a cell temperature of 300 K. Note that the current unit is different to that in Fig. 4.6

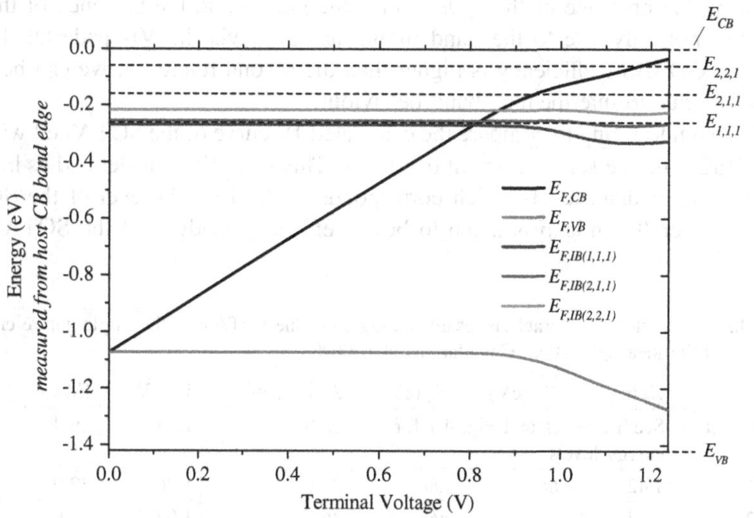

Fig. 4.8 The quasi Fermi levels of the three intermediate levels, as well as the VB and the CB, as a function of the bias voltage. The *horizontal dashed lines* show the energy levels of the three intermediate levels and the two bands. The calculations were made using the detailed balance model for the SOTA cell at 1000X concentration and a cell temperature of 300 K

4.5.2 JV Curve of the SOTA Cell: Comparison with Single-Gap Reference Cells

As has been discussed in the introduction to this book, the quantum dot solar cell has been proposed as a means of superseding the Shockley-Queisser limiting efficiency of single-gap solar cells. It is therefore interesting to compare the calculated JV curve of the SOTA cell with those of equivalent single-gap reference devices. Throughout this section, we shall make use of two reference IV curves. Both are for single gap solar cells and are calculated using the Shockley Queisser model [2, 16], which assumes the radiative limit and 100 % absorption of above-bandgap photons. The first reference is for a solar cell with $E_g = 1.42$ eV, which corresponds to GaAs. This is a suitable reference, since the studied QD-IBSC is based on introducing an intermediate band into a GaAs host, which in turn is sandwiched between a GaAs emitter and base. The second is for a solar cell with $E_g = 1.24$ eV, which is the effective VB-CB bandgap ($E_{g,eff}$) for the SOTA cell. The purpose for this choice of a second reference is explained in the following paragraph. The JV characteristics of both references are given in Table 4.3.

The effective bandgap of the QD-IBSC, which is reduced compared to bulk GaAs, could cause the QD-IBSC to have a higher efficiency than a GaAs reference, even in the absence of a contribution from the IB. This is because, for a single-bandgap device, a bandgap of $E_g = 1.24$ eV is better optimised for photovoltaic conversion (at 1000X) than a bandgap of $E_g = 1.42$ eV. In fact, this effect is exploited in quantum well solar cells, which purposefully tune the effective bandgap of a solar cell to better match the solar spectrum [25]. However, in this chapter, we are interested in increasing the solar cell efficiency due to intermediate band behaviour. We must therefore be sure that any projected increase in efficiency offered by the presence of the QDs is in some part due to the presence of the IB level and not only due to the band tuning achieved via the VB pedestal. If the projected QD-IBSC efficiency is higher than the second reference, we can be sure that this is due to intermediate band behaviour.

It is also interesting to compare the calculated JV curve of the SOTA cell with an ideal IBSC with the same principal bandgaps. This ideal IBSC is defined as having a single intermediate level, which corresponds to the IB(111) level of the SOTA cell. The overall bandgap is taken to be the effective bandgap of the SOTA cell.

Table 4.3 Calculated JV characteristics and bandgaps of the SOTA cell, the single-gap references devices and the ideal QD-IBSC. Calculations at 1000X

	E_g (eV)	E_H (eV)	E_L (eV)	J_{sc} (A cm^{-2})	V_{oc} (V)	Efficiency (%)
SOTA cell	See Table 4.1 and Fig. 4.1 for energy levels			28.9	1.18	29.4
Ref. 1	1.42	n/a	n/a	28.9	1.26	32.9
Ref. 2	1.24	n/a	n/a	36.7	1.07	34.0
Ideal IBSC	1.24	0.92	0.3	46.9	1.07	44.4

The JV curve for the ideal IBSC is calculated by applying the model in Ref. [1]; full absorption of incident photons and the radiative limit are assumed. The bandgaps and JV characteristics of the ideal IBSC are shown in Table 4.3.

The calculated JV curve of the SOTA cell is compared to those of the two single-gap references and the ideal QD-IBSC in Fig. 4.9. The JV characteristics are summarized in Table 4.3. As has been discussed previously, the extremely small contribution of the subbandgap current causes the SOTA cell to have a J_{sc} that is practically equal to that of Ref. 1 (which has the same bandgap as the GaAs host). The V_{oc} of the SOTA cell is lower than Ref. 1 due to radiative recombination via the VB pedestal and via the IB. Consequently, the efficiency (29.4 %) is below that of both references, even in the radiative limit at 1,000 suns concentration.

Clearly, for the SOTA cell to exceed the efficiency of both reference devices, it is necessary to increase the subbandgap photocurrent by increasing the photon absorption in the QDs. This is discussed in the following section. The J_{sc} of the ideal IBSC represents what could be achieved if the QDs absorbed all incident photons. The large difference between the ideal J_{sc} and the J_{sc} of the SOTA cell shows that there is a lot of scope to improve the efficiency via absorption enhancement.

In the radiative limit, the calculated V_{oc} of the SOTA cell is higher than that of Ref. 2 and of the ideal QD-IBSC. This is due to weak absorption, as has been discussed. The effect of increasing the absorption on the V_{oc} is also investigated in the following section.

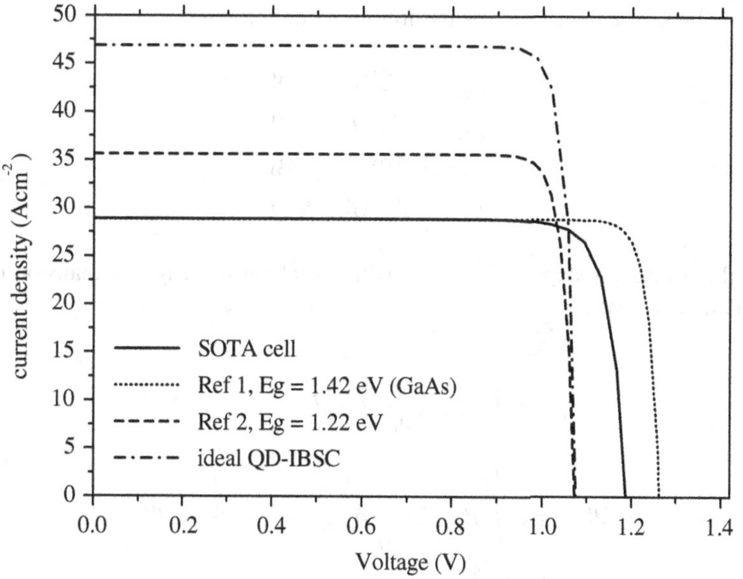

Fig. 4.9 JV curve of the SOTA cell in the radiative limit calculated using the detailed balance model. Also shown are JV curves for the two single-gap reference devices, as well as the ideal QD-IBSC, all of which are described in the text. Calculations at 1000X

4.6 Increasing the Subbandgap Absorption

Present QD solar cell prototypes exhibit an extremely low subbandgap photocurrent, which is one of the barriers preventing them from reaching high efficiencies. The present modelling work, described in this book and elsewhere [26], confirms that this low subbandgap photocurrent is due to weak absorption of subbandgap photons in the QD layer stack. For QD solar cells to achieve high efficiencies in the future, this subbandgap photocurrent will have to be increased dramatically. The purpose of this section is to investigate, by means of the detailed balance model, the effect that increasing the photon absorption has on the JV characteristics of the SOTA cell. Before discussing how absorption enhancement can be achieved, we must first adapt the detailed balance model to be able to deal with higher photon absorption. This section follows closely Ref. [26].

4.6.1 Correction of the Model to Account for Higher Absorption

In the initial derivation of the net current between two levels (Sect. 4.2.1), the exponential function in Eq. (4.9) was expanded in a Taylor series up to the first order. This is justified in the limit of low absorption. Here, the derivation is modified to allow for higher absorptions.

It will be convenient to make the following definitions:

$$\begin{aligned} \beta_{q,\varepsilon} &= \alpha_{q,\varepsilon}^{\max}(f_l - f_u)W \\ \beta_{x,y} &= \alpha_{x,y}^{\max}(f_l - f_u)W \\ \beta_z &= \alpha_z^{\max}(f_l - f_u)W \\ \beta_{iso} &= \alpha_{iso}^{\max}(f_l - f_u)W \end{aligned} \tag{4.34}$$

Working through steps (4.9)–(4.13) without taking the Taylor expansion of the exponential, we have

$$\begin{aligned} J_{l\to u}^{L} &= e\int_0^{E_g} dE_{phot}\frac{c}{n_{ref}}\rho_{phot}I^{L} \\ J_{l\to u}^{D} &= e\int_0^{E_g} dE_{phot}\frac{c}{n_{ref}}\rho_{phot}\left(f_{B,\mu T_c} - f_{B,0T_c}\right)I^{D} \end{aligned} \tag{4.35}$$

where I^L and I^D represent the following integrals

$$I^L = \int_0^{2\pi}\int_0^{\theta_c} d\varphi d\theta \cos\theta \sin\theta \left(1 - \exp\left[-\frac{\langle\beta_{q,\varepsilon}\rangle_\varepsilon}{\cos\theta}\right]\right) N_{inc}$$

$$I^D = \begin{bmatrix} \int_0^{2\pi}\int_0^{\theta_c} d\varphi d\theta \cos\theta \sin\theta \left(1 - 2\exp\left[-\frac{\langle\beta_{q,\varepsilon}\rangle_\varepsilon}{\cos\theta}\right] + \exp\left[-\frac{2\langle\beta_{q,\varepsilon}\rangle_\varepsilon}{\cos\theta}\right]\right) \\ + \int_0^{2\pi}\int_0^{\pi/2} d\varphi d\theta \cos\theta \sin\theta \left(1 - \exp\left[-\frac{2\langle\beta_{q,\varepsilon}\rangle_\varepsilon}{\cos\theta}\right]\right) \end{bmatrix} \quad (4.36)$$

Returning to Eq. (4.19), we observe that the QDs in the SOTA cell have xy symmetry, such that $\alpha_x^{max} = \alpha_y^{max} = \alpha_{x,y}^{max}$. Equation (4.19) then becomes

$$\left\langle \alpha_{\mathbf{k},\varepsilon}^{\max} \right\rangle_\varepsilon = \left(1 - \frac{\sin^2\theta_k}{2}\right)\alpha_{x,y}^{\max} + \frac{\sin^2\theta_k}{2}\alpha_z^{\max} - \cos\varphi_k \sin\varphi_k \sin^2\theta_k \alpha_{x,y}^{\max} - (\sin\varphi_k + \cos\varphi_k)\cos\theta_k \sin\theta_k \sqrt{\alpha_z^{\max}\alpha_{x,y}^{\max}} \quad (4.37)$$

Clearly, the many sine and cosine terms in $\left\langle \alpha_{\mathbf{k},\varepsilon}^{\max} \right\rangle_\varepsilon$ make the integrals I^L and I^D very difficult to compute. For I^L, we can take a Taylor expansion of the bracketed expression up to an arbitrary order. We observe that all terms but one in the integrand are then proportional to sin θ to the second power or higher. Making the approximation that these terms integrate to zero over the interval $0 \leq \theta \leq \theta_c$, the integral becomes

$$I^L = \int_0^{2\pi}\int_0^{\theta_c} d\varphi d\theta \sin\theta \left(1 - \exp\left[-\beta_{x,y}\right]\right) N_{inc} \quad (4.38)$$

where we have also made the approximation $\cos\theta \approx 1$, which is reasonable throughout the interval of integration. The illumination current for the transition is therefore

$$J_{l\to u}^L = q_e \int_0^{E_g} dE_{phot}\Phi_{inc}\left(1 - \exp\left[-\beta_{x,y}\right]\right) \quad (4.39)$$

as we would expect.

For I^D we can make a two approximations, both of which are justified numerically further on. Firstly, we take the first integral in the expression for I^D (Eq. 4.36) to be negligible compared to the second integral. Secondly, we perform the second integral taking $\langle \beta_{q,\varepsilon} \rangle_\varepsilon$ to be simply the mean value of β_x^{max}, β_y^{max} and β_z^{max}. Under these approximations, I^D becomes

$$I^D(\beta_{iso}) = \left[\int_0^{2\pi} \int_0^{\pi/2} d\varphi d\theta \cos\theta \sin\theta \left(1 - \exp\left[-\frac{2\beta_{iso}}{\cos\theta}\right]\right) \right] \\ = \pi\left(1 - (1 - 2\beta_{iso}) \exp[-2\beta_{iso}] + 4\beta^2 Ei[-2\beta_{iso}]\right) \tag{4.40}$$

where Ei[x] is the exponential integral defined by

$$\mathrm{Ei}[x] = -\int_{-x}^{\infty} dt \frac{\mathrm{e}^{-t}}{t} \tag{4.41}$$

To justify the approximations made in the expression of I^D, Fig. 4.10 shows the relative difference between the approximate I^D expressed in Eq. (4.40) and the exact I^D expressed in Eq. (4.36) for a range of values of $\beta_{x,y}$ and β_z. The exact I^D is calculated numerically. The relative error is in most cases below 10 %, but approaches almost 30 % if the absorption is much stronger for photons propagating in the z direction than in the xy plane. In all cases, the error is tolerable for the semi-quantitative analysis performed here.

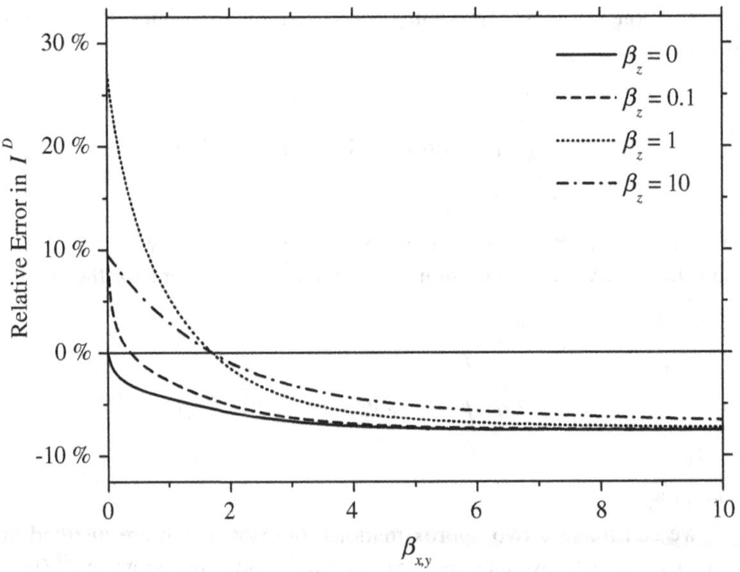

Fig. 4.10 Relative error induced by using approximate I^D expressed in Eq. (4.40) instead of the exact I^D expressed in Eq. (4.36), for a range of values of $\beta_{x,y}$ and β_z

Computationally, it is far preferable to use the approximate I^D. The exact I^D requires a numerical integral that would have to be executed in every iteration of the numerical solver used to find the QFLs. The approximate I^D is not quite analytical, since it contains the exponential integral Ei[x]; however, software packages such as Mathematica or Matlab will calculate its value using a truncated series, which is much faster than performing a numerical integration.

The dark current can now be expressed as

$$J^D_{l\to u} = q_e \int_0^{E_g} dE_{phot} \frac{2\pi n^2 \left(E_{phot}\right)^2}{h^3 c^2} \left(f_{B,\mu T_c} - f_{B,0T_c}\right) I^D\left(\beta_{iso}\right) \tag{4.42}$$

4.6.2 *Means of Increasing the Subbandgap Photon Absorption*

Two means of increasing the absorption are investigated. The first is by increasing the number of QDs in the QD stack. The second is by so-called light trapping, in which optical texturing of the substrate causes incident photons to be more strongly absorbed in the QD stack. Both means are described in the following paragraphs, along with the effects they have on the subbandgap photocurrents.

4.6.2.1 Increasing the Number of Quantum Dots

The overall subbandgap absorption can be increased by increasing the overall QD density per cm^2 of solar cell area. This can be achieved by increasing the areal QD density per layer and/or by increasing the number of QD layers. The density per cm^3 is also increased if we reduce the separation between layers so permitting to reduce the IB thickness. Our studies so far have focused on the SOTA cell, which is the sample labelled SB in Ref. [3]. This sample is a layer stack of 30 QD layers, each layer having an areal QD density of 4×10^{10} cm^{-2}. More recently, QD stacks have been grown with more layers and other have been grown with higher areal QD densities per layer. Reference [27] reported an InAs/GaNAs QD stack with 100 layers and areal QD density of 4×10^{10} cm^{-2}, and Ref. [28] reported an InAs/GaNAs QD stack with 300 layers and areal QD density of 6×10^{10} cm^{-2}. These represent an increase in the number of layers of a factor of 3 and 10, respectively, compared to the prototypes studied here. Other authors have reported GaAsSb QDs grown on (001) oriented GaAs with an areal density per layer of 4×10^{11} cm^{-2} [29], an increase by a factor of 10 compared to the prototypes studied here. In that work, only three QD layers were grown, making the overall density similar to our prototype, but the areal density per layer is remarkable.

Growing a high number of layers and achieving a high QD density per layer simultaneously will undoubtedly require further development. However, based on these results in the literature we can envisage the combination of a 10 fold increase in areal QD density per layer and a 10 fold increase in the number layers, leading to an overall increase in the number of QDs by a factor of 100. Further increases can be expected beyond this as technology develops.

4.6.2.2 Absorption Enhancement via Light Trapping

The probability that an incident photon is absorbed in a solar cell is proportional to the distance it travels in the absorber layer. This distance can be increased without increasing the layer thickness by employing so-called *light trapping* or *light confinement* techniques. These involve the incorporation of optical components into the solar cell with the aim of increasing the optical path length of incident photons within the absorber layer. The most basic light trapping technique is to attach a metal reflector to the rear of the solar cell. Photons not absorbed on making a single pass of the absorber layer are reflected and make another pass before leaving the cell, thus increasing their probability of absorption.

Better light trapping can be achieved by the use of textured surfaces. This was first proposed by Redfield in [30]. The principle is shown in Fig. 4.11. A textured surface deflects incident light into oblique directions. These have higher path lengths in the absorber layer than light which traverses normally. What's more, light deflected at certain angles can be confined in the absorber layer by total

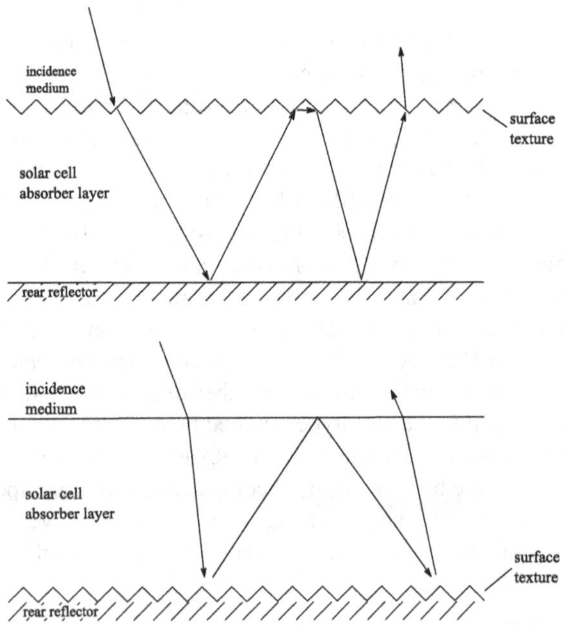

Fig. 4.11 Light-trapping property of textured surfaces. *Top* front surface texture. *Bottom* rear surface texture

internal reflection at the front surface, causing it to make multiple passes of the absorber layer before escape. Either, the front, rear, or both surfaces of the absorber layer can be textured to achieve this effect.

Two main texture types are used for light trapping in present solar cell designs. One is sub-micron scale surface roughening (Fig. 4.12, left). The other is geometric pattering of the surface with features whose dimensions are several microns (Fig. 4.12, right). The photons of interest for solar energy have wavelengths close to or below one micron. Hence, these textures scatter light according to geometric-optical phenomena, namely refraction and reflection from the facets of the textures. Sub-micron scale surface roughening tends to be favoured in thin-film solar cells, whereas geometric surface texturing is favoured in bulk c-Si solar cells.

The aim of both sub-micron scale surface roughening and geometric surface texturing is to achieve an isotropic scattering of incident photons inside the solar cell absorber layer. It has been shown that, in the limit of completely isotropic—or Lambertian—scattering, the mean optical path length enhancement of incident photons approaches $4n_{ref}^2$: the so called Lambertian limit [31]. This limit is valid for solar cells whose thickness is much larger than the wavelength of the incident photons. In the case of the SOTA cell, the absorber layer for subbandgap photons is the QD stack, which is a few microns thick. However, the QD layer stack is grown on a thick GaAs wafer substrate. If we assume that the reflector is placed on the rear of the substrate, then the geometric-optical picture sketched in Fig. 4.11, along with the Lambertian limit, is valid. For a GaAs substrate and host, with a refractive index of around $n = 3.5$, the Lambertian limit is a factor of around 49. In practise, more light tends be transmitted or reflected secularly than is scattered. Consequently, the surface textures used in present solar cells are believed to offer below-Lambertian light trapping, with optical path length enhancements of around 10 [32]. It should be observed that the Lambertian Limit is not an upper limit; it can be superseded at some wavelengths through the use of diffraction gratings [33], angle-selective filters [34] and near-field absorption enhancement techniques [35]. However, in this work, we take an optical absorption enhancement of 50 to be optimistic and absorption enhancements of 10 to be realistic, and do not consider the wavelength dependence of the absorption enhancement.

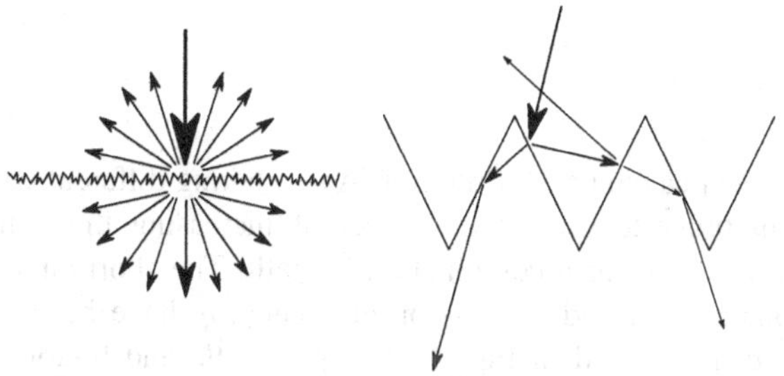

Fig. 4.12 *Left* light scattering by a sub-micron scale roughened surface. *Right* light scattering by geometric textures whose dimensions of many microns

4.6.2.3 The Effect of Absorption Enhancement on the Subbandgap Transition Currents

The two means of increasing the subbandgap photon absorption, described above, affect the subbandgap transition currents in different ways. The effect of increasing the number of QDs is to increase the absorbance ($\beta_{x,y}$, β_z and β) by a constant factor, which we denote Λ_{QD}. This is true whether the number of QDs is increased by increasing the number of layers or by increasing the areal QD density per layer. Note that increasing the number of QDs therefore affects both the illumination and dark currents. The effect of employing light trapping is quite different. Photons incident from the sun are scattered by the surface texture, making them more isotropic, and thus increasing their optical path length, inside the absorber layer. However, thermal and luminescent photons are already isotropic before reaching the surface texture. Light trapping is therefore expected to affect the illumination current, but not the dark current. To summarize, the detailed balance model is adapted to incorporate the effect of absorption enhancement by using the following expressions for the illumination and dark currents for each subbandgap transition.

$$
\begin{aligned}
J_{l\rightarrow u}^{L} &= q_e \int_0^{E_g} dE_{phot}\Phi_{inc}\left(1-\exp\left[-\Lambda_{opt}\Lambda_{QD}\beta_{x,y}\right]\right) \\
J_{l\rightarrow u}^{D} &= q_e \int_0^{E_g} dE_{phot}\frac{2\pi n^2\left(E_{phot}\right)^2}{h^3c^2}\left(f_{B,\mu T_c}-f_{B,0T_c}\right)I^D\left(\Lambda_{QD}\beta\right)
\end{aligned}
\tag{4.43}
$$

where Λ_{QD} is the factor by which the overall QD density has been increased per cm^2 of solar cell area (not per QD layer) with respect to the SOTA cell, and Λ_{opt} is the mean optical path length enhancement of incident photons achieved by light trapping. Note that Λ_{opt} only appears in J^L, whereas Λ_{QD} appears in both. From this insight, it should already be clear that absorption enhancement via light trapping will lead to a better efficiency boost than via increasing the number of QDs. This is examined quantitatively further on.

4.6.3 Results

JV curves have been calculated for the SOTA cell taking different combinations of Λ_{opt} and Λ_{QD} in order to predict the effect of increasing the number of QDs, employing light trapping, or a combination of both. The short circuit current, J_{sc}, open circuit voltage, V_{oc}, and conversion efficiency, η, have been extracted from these curves, and are plotted in Fig. 4.13 (top, middle and bottom respectively). Each curve is plotted as a function of the QD number enhancement factor, Λ_{QD}, and different curves are plotted for different values of the optical path length enhancement due to light trapping, Λ_{opt}. Results are plotted for $\Lambda_{opt} = 1$, $\Lambda_{opt} = 10$, and $\Lambda_{opt} = 50$, which respectively represent no light trapping, realistic light trapping using presently employed surface textures, and light trapping at the Lambertian limit. In each graph, the horizontal lines represent the respective quantity for the two single-gap references, as well as for the ideal QD-IBSC described previously.

Looking at the J_{sc}, we can see that, without light trapping, the number of QDs must be increased by a factor of around 300 for the J_{sc} of the SOTA cell to exceed that of Ref. 2. If absorption enhancements by factors of 10 and 50 are achieved via light trapping, this value drops to 30 and 5 respectively.

The behaviour of the V_{oc} requires careful discussion. Firstly, the model predicts that, under radiative conditions, the V_{oc} of the QD-IBSC with no absorption enhancement is higher than that of the ideal QD-IBSC. This appears to be a confusing result at first. The chosen ideal IBSC has an overall bandgap of $E_g = 1.24$ eV; this being equal to the effective bandgap, $E_{g,eff}$, of the QD-IBSC. It is generally supposed that the QD-IBSC's V_{oc} is limited at most by its effective bandgap, or, perhaps more likely, by one of the lower subbandgap transition energies ($E_{IB(111)}$, $E_{IB(121)}$ or $E_{IB(221)}$ in Fig. 4.1). We would therefore expect the V_{oc} of the QD-IBSC to be lower or equal to that of the ideal IBSC. However, since the subbandgap photon absorption/emission is so weak, and radiative conditions are assumed, the recombination via subbandgap transitions (those from the CB to the VB pedestal included) is extremely weak. This is to be compared to the above-bandgap photocurrent generated in the GaAs emitter (around 29 A cm^{-2}), which absorbs all incident photons above the GaAs bandgap (1.42 eV). Hence, in the absence of absorption enhancement, and in the radiative limit, the V_{oc} of the QD-IBSC is more similar to that of a single-gap GaAs device (around 1.42 V under the assumptions of the SQ model). Once again, the reader is reminded that present QD-IBSCs are dominated by non-radiative recombination, and therefore have much lower V_{oc}s (around 0.9 V for the prototype on which this study is based at 300 K and 1000X [36]). It should be mentioned that the fact that a high V_{oc} is predicted for a weakly absorbing solar cell is of no practical use. A highly efficient solar cell must absorb most incident photons; hence, only the results for stronger absorption are of practical importance.

We can see that there is a steady drop in the QD-IBSC's V_{oc} as the number of QDs increases. This follows from our discussion in the previous paragraph. As the number of QDs is increased, the total current becomes more strongly influenced by the subbandgap recombination currents, and the overall V_{oc} falls to a lower value. Importantly, the V_{oc} is predicted to fall to a value below that of Ref. 2 and the ideal QD-IBSC, meaning that the V_{oc} is limited not by the effective bandgap, but by one of the lower energy gaps (E_{t1}, E_{t2} or E_{t3} in Fig. 4.1). This is due to the CB QFL being pinned to one of the IB levels at open circuit at room temperature.

We can confirm from the calculations that, for high values of Λ_{QD}, the CB QFL is at around 0.16 eV below the CB onset, meaning that it is pinned at the IB(211) level. This follows naturally from the result in Sect. 4.4 and Refs. [33, 34], in which it was shown that there is a strong radiative thermal escape from the IB levels to the CB, but that this radiative escape always occurs via the IB(211) level. Since the CB

The V_{oc} is mainly affected by the dark current, and therefore does not vary significantly with optical path length enhancement, although there is a slight improvement of the V_{oc} for higher Λ_{opt}. This is probably related to the increased J_{sc}, which is analogous to the way in which the V_{oc} of a solar cell tends to improve logarithmically with increased concentration.

The most important metric resulting from the *JV* curve is the conversion efficiency. In the absence of light trapping ($\Lambda_{opt} = 1$), the predicted efficiency actually drops on increasing Λ_{QD} from 1 to 10; this is due to the predicted drop in V_{oc}. The efficiency then rises on increasing Λ_{QD} and surpasses the Ref. 1 and Ref. 2 for enhancements of $\Lambda_{QD} > 600$ and $\Lambda_{QD} > 800$ respectively (technologists must keep in mind a possible minimum of efficiency). On further increasing Λ_{QD}, the efficiency saturates at about 40 %, which is between that of the Ref. 2 and the ideal IBSC. This follows from our previous assertion that the V_{oc} is limited by E_{t2} and not E_{t1} or $E_{g,eff}$.

Clearly, an increase in the number of QDs by a factor of 800 or more is far beyond present technological capabilities. However, we can see from Fig. 4.13 (bottom) that, by employing light trapping, a much lower increase in the number of QDs are required for the QD-IBSC to supersede the efficiency of the single-gap references. For optical enhancements of $\Lambda_{opt} = 10$ and $\Lambda_{opt} = 50$, an increase in the number of QDs by factors of $\Lambda_{QD} = 30$ and $\Lambda_{QD} = 3$ are required respectively.

Comparing Fig. 4.13 top and bottom, we can see that absorption enhancement due to light trapping has a more profound effect on the efficiency than on the J_{sc}. To illustrate this, *JV* curves have been calculated for different combinations of Λ_{QD} and Λ_{opt} that lead to an overall absorption enhancement of 1000—the overall absorption enhancement being the product $\Lambda_{QD} \times \Lambda_{opt}$. The combinations are ($\Lambda_{opt} = 1$, $\Lambda_{QD} = 1000$), ($\Lambda_{opt} = 10$, $\Lambda_{QD} = 100$) and ($\Lambda_{opt} = 50$, $\Lambda_{QD} = 20$). The three *JV* curves are plotted in Fig. 4.14. All have the same J_{sc}, as would be expected. However, the V_{oc} is higher when a greater proportion of the enhancement is achieved by light trapping. Hence, the advantage of light trapping is two-fold; it reduces the number of QDs that are required for a high J_{sc}, making the QD stack more technologically feasible, and it also allows this high J_{sc} to be accompanied by a higher V_{oc} than would possible if no light trapping were employed.

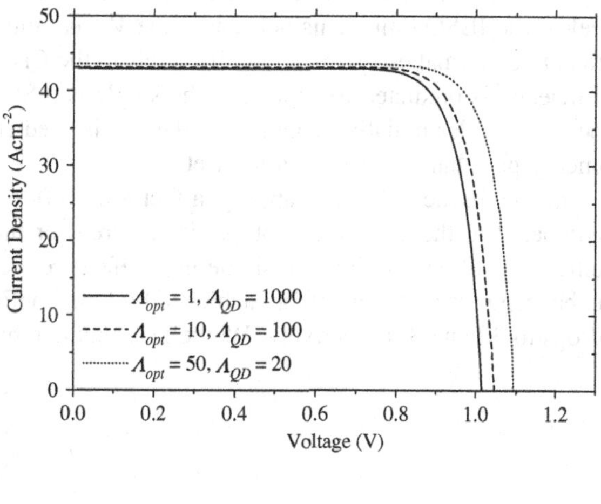

Fig. 4.14 JV curves for different combinations of E_{QD} and E_{opt}. All three combinations give a total absorption enhancement of $E_{QD} \times E_{opt} = 1{,}000$

4.7 Concluding Remarks of This Chapter

A non-idealized photon-electron detailed balance model has been developed to calculate the realistic intrasubband and intersubband transition currents in a real QD-IBSC prototype in the radiative limit. Instead of assuming absolute photon absorption, as is done in idealized detailed balance models, the model assumes realistic absorption, taking as input the optical matrix elements calculated in Chaps. 2 and 3 and in Refs. [5–8].

The most convincingly reproduced result is the temperature decay of the first IQE peak (interpreted as the VB → IB(111) transition). Two regimes are disclosed: at room temperature, the IQE is limited by the VB → IB transitions pumped by the monochromator, and at temperatures below 150–200 K, the IQE is limited by thermal escape from the IB to the CB.

The results confirm quantitatively that the measured thermal escape from the QD confined levels [3, 4] can be entirely accounted for by the absorption of thermal photons, incident from the surroundings and from the substrate, in a sequential many-photon absorption/emission process involving the ladder of QD excited states between the QD ground state and the CB. This mechanism was proposed in Ref. [20].

The results also reveal the sub-bandgap photogeneration is in fact a sequential three-photon process requiring photons with energy as low as 100 meV for the lowest energy transition. Most optical systems are opaque to photons of this energy (which correspond to a wavelength of 12.4 μm). This could explain the difficulty of measuring the intrasubband transitions in the past.

The *JV* characteristics of an InAs/GaAs QD-IBSC have been calculated using the realistic detailed balance model. The model is used to predict the J_{sc}, V_{oc} and overall efficiency that would be expected if non-radiative processes were suppressed and photon absorption were enhanced by different degrees, either by increasing the number of quantum dots, employing light trapping or a combination of the two. Calculations are made assuming 1000X sunlight concentration and 300 K operating temperature. Results are compared to the Shockley-Queisser limits for single-gap devices under the same conditions, as well as to the limiting efficiency of an ideal IBSC with the same principal bandgaps as the studied prototype.

On increasing the number of QDs indefinitely, the efficiency saturates at a value below the IBSC limit. This is because the V_{oc} is limited due to the aforementioned radiative thermal escape from the IB levels to the CB [22]. However, the QD-IBSC efficiency is predicted to supersede the single-gap Shockley-Queisser limit. This is because, under radiative conditions, the V_{oc} is predicted to be limited by a higher energy gap than the absorption onset.

In the absence of light trapping, a factor of 800-increase in the number of QDs, compared to the studied prototype, is required for the QD-IBSC to supersede the efficiency of an equivalent single-gap reference device. If optical path length enhancements of 10 and 50 are achieved by light confinement techniques, this value drops to 30 and 3 respectively. We believe that combing an increase in the number of QDs by a factor of 30 with an optical absorption enhancement of 10 is within present capabilities, or at least those expected in the near future.

Finally, it has been argued that absorption enhancement via light trapping is especially beneficial since it is expected to affect only the illumination current and not the dark current. Of course, both light trapping and an increase in the number of QDs will be required.

References

1. Luque A, Martí A (1997) Increasing the efficiency of ideal solar cells by photon induced transitions at intermediate levels. Phys Rev Lett 78(26):5014–5017
2. Shockley W, Queisser HJ (1961) Detailed balance limit of efficiency of p-n junction solar cells. J Appl Phys 32(3):510–519
3. Antolín E, Marti A, Farmer CD, Linares PG, Hernandez E, Sanchez AM, Ben T, Molina SI, Stanley CR, Luque A (2010) Reducing carrier escape in the InAs/GaAs quantum dot intermediate band solar cell. J Appl Phys 108(6):064513–064517
4. Antolín E, Marti A, Linares PG, Ramiro I, Hernandez E, Farmer CD, Stanley CR, Luque A (2010) Advances in quantum dot intermediate band solar cells. In: Paper presented at the 35th IEEE photovoltaic specialists conference (PVSC), Honolulu, 20–25 June 2010
5. Luque A, Marti A, Antolin E, Garcia-Linares P (2010) Intraband absorption for normal illumination in quantum dot intermediate band solar cells. Sol Energy Mater Sol Cells 94 (12):2032–2035
6. Luque A, Martí A, Mellor A, Fuertes Marrón D, Tobías I, Antolín E (2013) Absorption coefficient for the intraband transitions in quantum dot materials. Prog Photovoltaics: Res Appl 21:658–667. doi:10.1002/pip.1250
7. Luque A, Martí A, Antolín E, Linares PG, Tobías I, Ramiro I, Hernandez E (2011) New Hamiltonian for a better understanding of the quantum dot intermediate band solar cells. Sol Energy Mater Sol Cells 95(8):2095–2101
8. Luque A, Mellor A, Antolín E, Linares PG, Ramiro I, Tobías I, Martí A (2012) Symmetry considerations in the empirical k.p Hamiltonian for the study of intermediate band solar cells. Sol Energy Mater Sol Cells 103:171–183
9. Luque A, Mellor A, Ramiro I, Antolín E, Tobías I, Martí A (2013) Interband absorption of photons by extended states in intermediate band solar cells. Solar Energy Mater Solar Cells 115:138–144. doi:http://dx.doi.org/10.1016/j.solmat.2013.03.008
10. Mellor A, Luque A, Tobías I, Martí A (2014) Realistic detailed balance study of the quantum efficiency of quantum dot solar cells. Adv Funct Mater. 24(3):339–345. doi:10.1002/adfm.201301513
11. Luque A, Mellor A, Tobías I, Antolín E, Linares PG, Ramiro I, Martí A (2013) Virtual-bound, filamentary and layered states in a box-shaped quantum dot of square potential form the exact numerical solution of the effective mass Schrödinger equation. Phys B: Condens Matter 413 (0):73–81. doi:http://dx.doi.org/10.1016/j.physb.2012.12.047
12. Messiah A (1962) Quantum mechanics, vol 2. North Holland Publishing Company, Amsterdam
13. Wurfel P (1982) The chemical potential of radiation. J Phys C: Solid State Phys 15(18):3967
14. Luque A, Marti A (2010) On the partial filling of the intermediate band in IB solar cells. IEEE Trans Electron Devices 57(6):1201–1207. doi:10.1109/ted.2010.2045681
15. Luque A, Linares PG, Antolín E, Ramiro I, Farmer CD, Hernández E, Tobías I, Stanley CR, Martí A (2012) Understanding the operation of quantum dot intermediate band solar cells. J Appl Phys 111:044502. doi:10.1063/1.3684968

16. Luque A, Hegedus S (eds) (2003) Handbook of photovoltaic science and engineering. Wiley, Chichester
17. Luque A, Marti A, Cuadra L (2001) Thermodynamic consistency of sub-bandgap absorbing solar cell proposals. IEEE Trans Electron Devices 48(9):2118–2124. doi:10.1109/16.944204
18. Ramiro I, Antolin E, Steer MJ, Linares PG, Hernandez E, Artacho I, Lopez E, Ben T, Ripalda JM, Molina SI, Briones F, Stanley CR, Marti A, Luque A (2012) InAs/AlGaAs quantum dot intermediate band solar cells with enlarged sub-bandgaps. In: Photovoltaic specialists conference (PVSC), 2012 38th IEEE, pp 000652–000656. 3–8 June 2012
19. Antolín E, Martí A, Luque A (2011) The lead salt quantum dot intermediate band solar cell. In: Photovoltaic specialists conference (PVSC), 37th IEEE, pp 1907–1912. doi:10.1109/pvsc.2011.6186324
20. Luque A, Marti A, Antolin E, Linares PG, Tobias I, Ramiro I (2011) Radiative thermal escape in intermediate band solar cells. AIP Adv 1(2):022125–022126
21. Linares PG, Marti A, Antolin E, Luque A (2011) III-V compound semiconductor screening for implementing quantum dot intermediate band solar cells. J Appl Phys 109(1):014313–014318
22. Mellor A, Luque A, Tobias I, Marti A (2012) The influence of quantum dot size on the sub-bandgap intraband photocurrent in intermediate band solar cells. Appl Phys Lett 101 (13):133909–133904
23. Bastard G, Ziemelis UO, Delalande C, Voos M, Gossard AC, Wiegmann W (1984) Bound and virtual bound states in semiconductor quantum wells. Solid State Commun 49(7):671–674
24. Bailey CG, Forbes DV, Polly SJ, Bittner ZS, Dai Y, Mackos C, Raffaelle RP, Hubbard SM (2012) Open-circuit voltage improvement of InAs/GaAs quantum-dot solar cells using reduced InAs coverage. IEEE J Photovoltaics 2(3):269–275. doi:10.1109/JPHOTOV.2012.2189047
25. Barnham KWJ, Duggan G (1990) A new approach to high-efficiency multi-band-gap solar-cells. J Appl Phys 67(7):3490–3493. doi:10.1063/1.345339
26. Mellor A, Luque A, Tobías I, Martí A (2014) The feasibility of high-efficiency InAs/GaA quantum dot intermediate band solar cells. Sol Energy Mater Sol Cells 130:225–233. doi:10.1016/j.solmat.2014.07.006
27. Takata A, Oshima R, Shoji Y, Akahane K, Okada Y (2010) Fabrication of 100 layer-stacked InAs/GaNAs strain-compensated quantum dots on GaAs (001) for application to intermediate band solar cell. In: Photovoltaic specialists conference (PVSC), 2010 35th IEEE, pp 001877–001880. 20–25 June 2010
28. Akahane K, Yamamoto N, Kawanishi T (2011) Fabrication of ultra-high-density InAs quantum dots using the strain-compensation technique. Phys Status Solidi (a) 208(2):425–428. doi:10.1002/pssa.201000432
29. Fujita H, Yamamoto K, Ohta J, Eguchi Y, Yamaguchi K (2011) In-plane quantum-dot superlattices of InAs on GaAsSb/GaAs(001) for intermediate band solar-cells. In: Photovoltaic specialists conference (PVSC), 2011 37th IEEE, pp 002612–002614. 19–24 June 2011
30. Redfield D (1974) Multiple-pass thin-film silicon solar cell. Appl Phys Lett 25(11):647–648
31. Yablonovitch E, Cody GD (1982) Intensity enhancement in textured optical sheets for solar cells. IEEE Trans Elect Dev 29(2):300
32. Nelson J (2003) The physics of solar cells. Imperial College Press, London
33. Mellor A, Tobías I, Martí A, Mendes MJ, Luque A (2011) Upper limits to absorption enhancement in thick solar cells using diffraction gratings. Prog Photovoltaics Res Appl 19 (6):676–687. doi:10.1002/pip.1086
34. Ulbrich C, Fahr S, Üpping J, Peters M, Kirchartz T, Rockstuhl C, Wehrspohn R, Gombert A, Lederer F, Rau U (2008) Directional selectivity and ultra-light-trapping in solar cells. Phys Status Solidi (a) 205(12):2831–2843. doi:10.1002/pssa.200880457
35. Yu Z, Raman A, Fan S (2010) Fundamental limit of nanophotonic light trapping in solar cells. Proc Nat Acad Sci USA 107:17491–17496. doi:10.1073/pnas.1008296107
36. Linares PG, Martí A, Antolín E, Ramiro I, Luque A (2011) Voltage recovery in intermediate band solar cells. Sol Energy Mater Sol Cells 98:240–244

Chapter 5
Interband Optical Absorption in Quantum Well Solar Cells

Abstract In quantum well solar cells, the eigenstates are bound in one direction and extended in the other two directions. The Empirical k·p Hamiltonian method can also be applied to this case. The envelopes of the four Γ-point Bloch functions contain a one-dimensional bound function in the direction of the growth of the quantum well layer multiplied by a two-dimensional plane wave in the plane perpendicular to the growth (horizontal). The envelope depends on the wave vector of the extended functions and is different for each Γ-point Bloch function. The dipole matrix of the optical transitions differs from the one used in preceding chapters in this book; the optical dipole operator previously used would be non-Hermitical when the initial and the final eigenfunctions are extended in some dimension. This is a very important aspect studied with detail in this chapter. The transitions caused by vertical photons conserve the horizontal wavevector; otherwise the matrix element is zero. An experimental quantum well solar cell has been modeled and its quantum efficiency has been simulated in reasonably good agreement with the measured curve. A clear description of the transitions produced is provided.

Keywords Solar cells · Quantum calculations · k·p methods · Quantum wells · Photon absorption

5.1 Introduction

Quantum wells (QWs) are widely used for numerous applications. In particular they are used as infrared photo detectors (QWIPs) [1]. In many of these applications, the energy spectrum of the QW is of interest only in the vicinity of the conduction band (CB), and the transitions of interest are the intraband photon absorption, which is the optical absorption between states detached from the CB. These problems may be reasonably modeled by the use of an effective mass and a potential that is the CB offset between the barrier material and the well material. However, this eludes the interpretation of interesting devices, such as solar cells whose bandgaps are tuned

A. Luque and A.V. Mellor, *Photon Absorption Models in Nanostructured Semiconductor Solar Cells and Devices*, SpringerBriefs in Applied Sciences and Technology, DOI 10.1007/978-3-319-14538-9_5

using QWs [2–4], because modeling of the absorption properties is beyond the reach of such a simple model. Approximate models for this absorption have also been used [5], but they are largely empiric so do not provide a deep insight into the interband mechanisms.

The EKPH [6, 7] has been developed to provide an easier explanation of the absorption properties of the quantum dot solar cells [8, 9], which is partly produced between the valence band (VB) and the *cb* states. This is referred to as interband absorption. This part presents the application of the EKPH to a better understanding of the wavefunctions associated to several of the bands in semiconductor QWs and to the interband photon absorption.

Even for intraband transitions, only within *cb* states, the absorption formulas in Eq. (2.70) are not applicable to QWs. The appropriate formulas will be developed in this chapter.

The QW studied in this section has been grown on a p-doped GaAs substrate, the solar cell consists of a GaAs n-i-p structure. Within the i-region, there are 40 repeats of 8 nm $In_{0.2}Ga_{0.8}As$ QWs, separated by 14.6 nm thick $GaAs_{0.81}P_{0.19}$ spacers to form the QWs; they are also useful for compensating the strain induced by the QWs. The QW-region is sandwiched between and n-doped emitter and a p-doped base. On top of the emitter there is an $Al_{0.5}In_{0.5}P$ window layer and below the base an InGaP back surface field layer, respectively, in order to minimize surface recombination. The layers forming the solar cells are reported in Table 5.1.

Table 5.1 Layer description of a quantum well solar cell prototype

Layer no.	Layer description	Material	Dopant	Doping (cm^{-3})	Thickness (nm)	Comments
11	Cap	GaAs	n	1e19	150.0	
10	Window	Al0.5InP	n	5.0E+18	30.0	
9	Emitter	GaAs	n	1.0E+18	200.0	
8	i-spacer	GaAs	–		5.0	
7	Half barrier	GaAsP0.19	–		7.3	
6	Quantum well	In0.20GaAs	–		8.0	40 repeats
5	Half barrier	GaAsP0.19	–		7.3	
4	i-spacer	GaAs	–		5.0	
3	Base	GaAs	p	8.0E+16	2000.0	
2	BSF	InGaP	p	1.0E+18	30.0	
1	Buffer layer	GaAs	p	2.0E+18	500.0	
0	Substrate	GaAs	p			

This chapter follows closely the lines of Ref. [10]

5.2 Envelope Functions

The calculation of the envelope functions follows the subsequent path [6, 7]: (1) resolution of the diagonalized TISE; (2) calculation of the solution's Fourier Transform; (3) multiplication by the appropriate element of the transformation matrix; (4) calculation of the Inverse Fourier Transform.

5.2.1 Solutions of the Time Independent Schrödinger Equation for the Diagonalized Hamiltonian

For a semiconductor with a QW the diagonalized TISE is

$$-\frac{\hbar^2}{2m^*}\nabla^2\Phi(r) + U(z)\Phi(r) = E\Phi(r) \tag{5.1}$$

We can separate variables and write

$$\Phi(\mathbf{r}) = \zeta(z)\exp\left(i(k'_x x + k'_y y)\right)\Big/\sqrt{L_x L_y} \tag{5.2}$$

where

$$-\frac{\hbar^2}{2m^*}\frac{d^2\zeta(z)}{dz^2} + U(z)\zeta(z) = E_z\zeta(z) \tag{5.3}$$

The asterisk refers to the effective mass in the band we are considering: the conduction band (*cb*) or the light hole (*lh*), heavy hole (*hh*) or split-off (*so*) VBs. Henceforth, CB will be used for states with energy above the CB edge of the barrier material and (*cb*) for states derived from CB states, even if they are within the host bandgap. The potential is the band offset (with changed sign for the VBs). The solution of this equation is usually easy. For square potential wells, it can be found in textbooks. The primes in the *k*'s refer to specific values of these variables, acting as parameters for the solution selected. The *x*- and *y*-dependent functions are normalized with the $\sqrt{L_x L_y}$ denominator. $\zeta(z)$ is also expected to be normalized.

The eigenenergy of an eigenfunction is

$$E = E_z + \frac{\hbar^2 k_x'^2}{2m^*} + \frac{\hbar^2 k_y'^2}{2m^*} \tag{5.4}$$

For values of E_z below the well rim, the *z*-solutions are bound. If the potential energy is a square well, $\zeta(z)$ can be even or odd. In both cases it is formed of a harmonic function (cosine or sine respectively) inside the well flanked by two fading exponentials outside it. Only selected values of E_z are valid (those making the rising exponential solutions disappear, whilst maintaining continuity of the wavefunction and its derivative at the well boundaries). For values above the rim, the solutions are extended [11]; here, we restrict our analyses to the case in which $\zeta(z)$ is bound: the only case leading to sub-bandgap absorption.

5.2.2 Calculating Fourier Transforms

The plane-wave development coefficients of $\Phi(\boldsymbol{r})$ in Eq. (5.2) can be directly written in terms of Fourier Transforms, as explained in Sect. 3.1.4. The DFT is only used for the z-function, requiring numerical calculations. For the x- and y-functions, we recall that $\mathscr{F}\left[e^{ik'_x x}\right](k_x) = \sqrt{2\pi}\delta(k'_x - k_x)$, therefore,

$$\begin{aligned}\phi(\boldsymbol{k}) &= \phi_z(k_z)\phi_x(k_x)\phi_y(k_y)\\ &= \frac{\sqrt{L_z}(2\pi)^2}{L_x L_y \sqrt{N}}\mathscr{F}\left[\{\zeta_{n_z}\}\right]_{\kappa_z}\delta(k'_x - k_x)\delta(k'_y - k_y)\end{aligned} \tag{5.5}$$

where all normalizing and conversion factors in Eqs. (3.20), (3.22) and (5.2) have been considered.

5.2.3 Calculating the Envelope Functions

To calculate the envelope functions, $\phi(\boldsymbol{k})$ is to be multiplied by one of the elements $T_\gamma^{(i,j)}(\boldsymbol{k})$ in Table 3.1. The row of the element, from 1–4, corresponds to the *cb, lh, hh* and *so* bands respectively. The column of the element, from 1–4 corresponds to the *X*-, *Y*-, *Z*- and *S*-envelopes respectively [7]. With the new plane wave development so obtained, $\Psi(\mathbf{r})$ is calculated as follows. The IFI (see Sect. 3.1.4 for nomenclature) is calculated with respect to the variables x and y and finally the IDFT is calculated with respect to the variable z. Starting with the variables x and y we obtain

$$\mathscr{F}^{-1}\left[{}^{\gamma}T_{i,j}(k_x,k_y)\phi(k_x,k_y)\right](x,y) = \frac{\sqrt{L_z}}{\sqrt{N}}{}^{\gamma}T_{i,j}(k'_x,k'_y,k_z)\mathscr{F}\left[\{\zeta_{n_z}\}\right]_{\kappa_z}\frac{e^{i(k'_x x + k'_y y)}}{\sqrt{L_x L_y}} \tag{5.6}$$

Here, the nomenclature shows that we deal with a two-dimensional inverse Fourier Integral in which the variables of the transformation are $\{k_x, k_y\} \rightarrow \{x, y\}$. This is obtained by successively performing the one-dimensional transformations $k_x \rightarrow x$ and $k_y \rightarrow y$. $T_\gamma^{(i,j)}(\boldsymbol{k})$ and $\phi(\boldsymbol{k})$ are also functions of k_z, but in this transformation process k_z is considered a parameter. Next, the IDFT is performed with respect to κ_z. A rather simple expression is found for the z-function:

$$\zeta_{i,j}(n_z L_z/N) = \mathscr{F}^{-1}\left[\left\{{}^{\gamma}T_{i,j}(k_x', k_y', 2\pi\kappa_z/L_z)\mathscr{F}\left[\{\zeta_{n_z}\}\right]_{\kappa_z}\right\}\right]_{n_z} \tag{5.7}$$

The full envelope functions are

$$^{qn}\Psi_{i,j}(\boldsymbol{k}) = {}^{qn}\zeta_{i,j}(z, k_x', k_y',)\frac{e^{i(k_x' x + k_y' y)}}{\sqrt{L_x L_y}} \tag{5.8}$$

Compared to the envelope Eq. (3.2), we have added to the envelope a second index representing the band (index i). A full description of the envelope requires an additional quantum number (qn) that defines the specific solution for the Φ we are considering within the band in which it has been solved.

Equation (5.8) shows that the x and y dependence of the envelope functions is a plane wave with the same wavenumbers as Φ.

These wavenumbers affect $^{qn}\zeta_{i,j}$ and, consequently, the projection of Φ onto the different GBFs (which is $\langle {}^{qn}\zeta_{i,j} | {}^{qn}\zeta_{i,j}\rangle$ [6] if $^{qn}\Psi_{i,j}$ is normalized to one).

Note that all the constants have disappeared from Eq. (5.7) resulting from the fact that direct and then inverse transforms have been performed, which are affected by reciprocal factors when converting them to the plane-wave developments. The constants would have disappeared with any type of Fourier Transforms.[1]

We present in Fig. 5.1 the *cb*, *hh* and *lh* projection of the ground states $^{1}\zeta_{cb}$, $^{1}\zeta_{hh}$ and $^{1}\zeta_{lh}$ on the different GBFs. This projection is the square of the envelope norm. Note that the main projection of the *cb* states is in the *S*-envelope which is weaker (or even totally absent for the *hh*) in the VB states.

5.3 Photon Absorption

5.3.1 Dipole Matrix Elements

Unlike what was stated in Sects. 2.1.4, 2.4 and 3.4, the dipole matrix element ruling the photon-electron interaction is $\langle \Xi | \epsilon \cdot \boldsymbol{p} | \Xi' \rangle$ [13, 14] (instead of $\langle \Xi | \epsilon \cdot \boldsymbol{r} | \Xi' \rangle$) where ϵ is the light-polarization vector and $\boldsymbol{p}$ the momentum operator (proportional to the gradient). This new expression of the dipole matrix element is required when both the initial and the final state are extended in, at least one and the same dimension. Further justification will be given later.

[1] Fourier transforms use different constants in the different fields of the science in which they are applied.

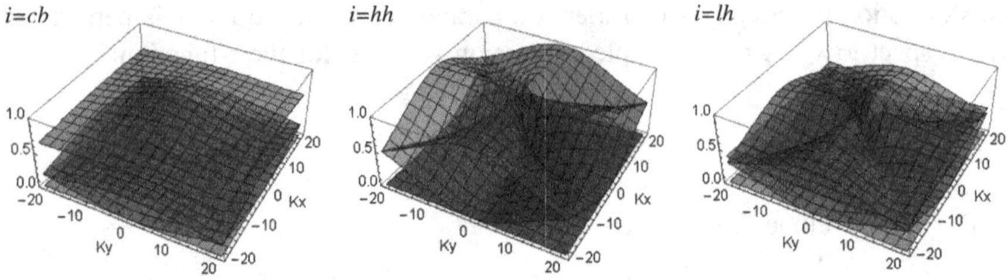

Fig. 5.1 Projection (the *square* of the envolvent norm) $\langle {}^{qn}\zeta_{i,j} | {}^{qn}\zeta_{i,j} \rangle$ of the *cb* ($i = cb$), *hh* ($i = hh$) and *lh* ($i = lh$) ground ($qn = 1$) states on the GBFs versus ($k_x d$, $k_y d$). *Blue*, *red*, *green* and *purple* for the *X*-,*Y*-, *Z*-, and *S*-GBFs (j = 1, 2, 3, 4) respectively. $d = 10$ nm. The projection depends on ($k_x d$, $k_y d$). Note that the main projection of the *cb* states is in the *S*-envelope which is weaker (or even totally absent for the *hh*) in the VB states. Reproduced with permission. © 2013, Elsevier [10]

$\boldsymbol{p}(u\Psi) = u\boldsymbol{p}\Psi + \Psi\boldsymbol{p}u$. Due to the different scale of distances associated to the k·p method, and applying the integral factorization rule of Eq. (2.1), $\langle u\Psi | \boldsymbol{\epsilon} \cdot \boldsymbol{p} | u'\Psi' \rangle \cong \langle u|u' \rangle \langle \Psi | \boldsymbol{\epsilon} \cdot \boldsymbol{p} | \Psi \rangle + \langle u | \boldsymbol{\epsilon} \cdot \boldsymbol{p} | u' \rangle \langle \Psi | \Psi' \rangle$ [15–17]. The *u* functions are average-normalized (see Eq. (2.13) for this concept) within the unit cells. Applying this and the multiband envelope function development of Eq. (3.2), we obtain

$$\begin{aligned} \langle \Xi | \boldsymbol{\epsilon} \cdot \boldsymbol{p} | \Xi' \rangle &= \sum_{v,v'} \langle u_{ov} | u_{0,v'} \rangle \langle \Psi_v | \boldsymbol{\epsilon} \cdot \boldsymbol{p} | \Psi'_{v'} \rangle + \sum_{v,v'} \langle u_{0,v} | \boldsymbol{\epsilon} \cdot \boldsymbol{p} | u_{0,v'} \rangle \langle \Psi_u \mid \Psi'_{u'} \rangle \\ &= \sum_{v} \langle \Psi_v | \boldsymbol{\epsilon} \cdot \boldsymbol{p} | \Psi'_{v'} \rangle + \sum_{v,v'} \langle u_{0,v} | \boldsymbol{\epsilon} \cdot \boldsymbol{p} | u_{0,v'} \rangle \langle \Psi_u \mid \Psi'_{u'} \rangle \end{aligned} \tag{5.9}$$

where we have considered that $\langle u_{0v} | u_{0v'} \rangle = \delta_{v,v'}$.

Any polarization orientation may be developed into its coordinate axes [7, 18] leading to terms containing p_x, p_y, and p_z. For the calculation of $\langle u_{0v} | \boldsymbol{\epsilon} \cdot \boldsymbol{p} | u_{0v'} \rangle$ we can use the symmetry properties of the *u*s in zincblend crystals. Looking at Eq. (3.16), the only non-null terms [6, 17] are $\langle S|p_x|X \rangle = \langle S|p_y|Y \rangle = \langle S|p_z|Z \rangle = P_0$ with P_0 to be found in Eq. (3.18). Thus, for the case of *x*-polarization, Eq. (5.9) becomes

$$\begin{aligned} \langle \Xi | p_x | \Xi' \rangle &= \sum_{v} \langle \Psi_v | p_x | \Psi'_v \rangle + P_0 \langle \Psi_S | \Psi'_X \rangle \equiv \Pi_x + P_0 \langle \Psi_S | \Psi'_X \rangle \\ \Pi_x &\equiv \langle \Psi_X | p_x | \Psi'_X \rangle + \langle \Psi_Y | p_x | \Psi'_Y \rangle + \langle \Psi_Z | p_x | \Psi'_Z \rangle + \langle \Psi_S | p_x | \Psi'_S \rangle \\ &\equiv \Pi_{xX} + \Pi_{xY} + \Pi_{xZ} + \Pi_{xS} \end{aligned} \tag{5.10}$$

Vertical illumination involves photons with horizontal polarization. Let us examine, for example, the Π_{xS} matrix element for a $hh \rightarrow cb$ transition. It is a function of the differences $k_{x,hh} - k_{x,cv} = \omega_x$ and $k_{y,hh} - k_{y,cv} = \omega_y$. The bound part of the *hh* and *cb* wavefunctions correspond respectively to the quantum numbers *qn* and *qn'*. Thus,

$$\begin{aligned} \Pi_{xS} &= \left\langle {}^{qn}\zeta_{cb,S} \mid {}^{qn'}\zeta_{hh,S} \right\rangle \int\limits_{-L_y/2}^{L_y/2} \frac{e^{i\omega_y y}}{L_y} dy \int\limits_{-L_x/2}^{L_x/2} \frac{e^{-ik_{cb,x}x} p_x e^{ik_{hh,x}x}}{L_x} dx \\ &= \left\langle {}^{qn}\zeta_{cb,S} \mid {}^{nq'}\zeta_{hh,S} \right\rangle \frac{\sin(\omega_y L_y/2)}{\omega_y L_y/2} \frac{\hbar k_{hh,x} \sin(\omega_x L_x/2)}{\omega_x L_x/2} \end{aligned} \tag{5.11}$$

The only permitted *k*s are integer multiples of $2\pi/L$, and the same for their differences ω. The *y*-function is one when $\omega_y \to 0$ and is strictly zero for the permitted lateral differences, of $\pm 2\pi/L$, $\pm 4\pi/L$, etc. Thus, k_y is conserved. The *x*-function is $\hbar k_{hh,x}$ for $\omega_x \to 0$ and zero for the permitted lateral values. Thus, k_x is also conserved. In summary, by dropping the band index from k_x,

$$\Pi_{xS} = \hbar k_x \left\langle {}^{qn}\zeta_{cb,S} \mid {}^{qn'}\zeta_{hh,S} \right\rangle \tag{5.12}$$

The same is applicable to other matrix elements. Therefore,

$$\begin{aligned} \Pi_x &= \hbar k_x \left\langle {}^{qn}\zeta_{cb} \mid {}^{qn'}\zeta_{hh} \right\rangle \quad \text{with} \\ \left\langle {}^{qn}\zeta_{cb} \mid {}^{qn'}\zeta_{hh} \right\rangle &\equiv \left\langle {}^{qn}\zeta_{cb,X} \mid {}^{qn'}\zeta_{hh,X} \right\rangle + \left\langle {}^{qn}\zeta_{cb,Y} \mid {}^{qn'}\zeta_{hh,Y} \right\rangle \\ &\quad + \left\langle {}^{qn}\zeta_{cb,Z} \mid {}^{qn'}\zeta_{hh,Z} \right\rangle + \left\langle {}^{qn}\zeta_{cb,S} \mid {}^{qn'}\zeta_{hh,S} \right\rangle \end{aligned} \tag{5.13}$$

(the envelope internal product without envelope subindex means, by definition, the sum of internal products).

$\left\langle {}^{qn}\Xi_{cb} | {}^{qn'}\Xi_{hh} \right\rangle$(= 0 because both functions are eigenvectors of the Hamiltonian of different energy) can be calculated using a procedure similar to this used in Eq. (5.11). It is found that $\left\langle {}^{qn}\zeta_{cb} | {}^{qn'}\zeta_{hh} \right\rangle = \left\langle {}^{qn}\Xi_{cb} | {}^{qn'}\Xi_{hh} \right\rangle = 0$, so that Eq. (5.10) can be rewritten as

$$\left\langle {}^{qn}\Xi_{cb} | v | {}^{qn'}\Xi_{hh} \right\rangle = P_0 \left\langle {}^{qn}\Psi_{cb,S} | {}^{qn'}\Psi_{hh,X} \right\rangle = P_0 \left\langle {}^{qn}\zeta_{cb,S} | {}^{qn'}\zeta_{hh,X} \right\rangle \tag{5.14}$$

A similar treatment is to be used for *y*-polarization. The *z*-polarization, only possible with non-vertical illumination, involves another procedure that is only briefly discussed. The reason is that, in this case, Π_z is not cancelled, and therefore it is better to use the theorem of Ehrenfest: $i\hbar \boldsymbol{p}/m = [\boldsymbol{r}, H_z]$ [13, 14] leading to an element of matrix proportional to the *z*-operator and not p_z. Since the cancellation of Π_x results from the plane wave feature of the *x*-wavefunctions, it is not found for quantum dots (QDs) either [6, 7] and the $\boldsymbol{r}$-version of the optical dipole operator is used along most of the chapters of this book (excepting this chapter and Chap. 6) and in particular is the one used in Sects. 2.1.4, 2.4 and 3.4. Taking this into account, for QWs, the matrix element is proportional to $P_0\left\langle {}^{qn}\zeta_{cb,S}|{}^{qn'}\zeta_{hh,X}\right\rangle$, while for QDs, it is proportional to $\left\langle \Psi_{cb,X}|x|\Psi'_{hh,X}\right\rangle + \left\langle \Psi_{cb,Y}|x|\Psi'_{hh,Y}\right\rangle +$ $\left\langle \Psi_{cb,Z}|x|\Psi'_{hh,Z}\right\rangle + \left\langle \Psi_{cb,S}|x|\Psi'_{hh,S}\right\rangle$. In QDs, the cb eigenfunctions are strongly projected onto the S-GBFs, but weakly onto the VB GBFs, while the opposite is true for the hh and lh eigenfunctions. Therefore, the integral overlaps are weak for the QDs but may be strong for the QWs. This might be interpreted as the cause of a stronger absorption in QWs; however, this is not completely true because the use the theorem of Ehrenfest introduces a factor in the above-cited sum of elements of matrix that is larger than in $P_0\left\langle {}^{qn}\zeta_{cb,S}|{}^{qn'}\zeta_{hh,X}\right\rangle$. Therefore, the main explanation of why the measured absorption coefficients are about two orders of magnitude stronger in QWs than in QDs is mainly to be found in the many states associated to the continuous spectrum of the horizontal part of the QW eigenfunctions.

The element of matrix, $\langle \Xi|\boldsymbol{\epsilon}\cdot\boldsymbol{p}|\Xi'\rangle$, used in this section is the one derived directly for quantum electrodynamics [14, 19]. The $i\hbar\boldsymbol{p}/m = [\boldsymbol{r}, H_z]$ transformation cannot be used when the initial and final wavefunctions are extended functions (as is the case for the x- and y-wavefunctions of the QWs) because $\boldsymbol{r}$ as an operator is not Hermitical: it does not fulfill the periodicity in the Ω volume. This is why we use here an element of matrix that differs from the one that is described in Sects. 2. 1.4, 2.4 and 3.4 and is used throughout this paper when at least one of the states (initial or final) is bound. It has to be said that the $\langle \Xi|\boldsymbol{\epsilon}\cdot\boldsymbol{p}|\Xi'\rangle$ element of matrix is more widely used by solid state physicists than $\langle \Xi|\boldsymbol{\epsilon}\cdot\boldsymbol{r}|\Xi'\rangle$ (they usually deal with extended states); our preference for the latter derives from its easier utilization in our context and from is more intuitive interpretation (once squared) as a capture section for photons, helpful for device engineers.

5.3.2 The Absorption Coefficient for x- or y-Polarization

As discussed in Sect. 2.4, the number of photon absorptions per unit of time [6, 13] when the electron passes from state $|\Xi\rangle$ to state $|\Xi'\rangle$ is

$$w_{\Xi\to\Xi'} = \Delta n_{ph}\frac{\pi e^2\hbar}{n_{ref}^2 m_0^2 \varepsilon_0 E}|\langle \Xi|\boldsymbol{\epsilon}\cdot\boldsymbol{p}|\Xi'\rangle|^2\delta(E_\Xi - E_{\Xi'} - E) \tag{5.15}$$

where ε_0 is the vacuum permittivity, n_{ref} is the refractive index of the medium and Δn_{ph} is the density of photons in all the modes of energy E. This density can be related to the photon flux by $\Delta F_{ph} = (c/n_{ref})\Delta n_{ph}$. Taking into account that the number of photon absorptions per unit of time is related to the elementary (per QW layer) absorption coefficient $\alpha^{layer}_{\Xi\to\Xi'}$ in the transition under consideration by $w_{\Xi\to\Xi'} = \Delta F_{ph}\alpha^{layer}_{\Xi\to\Xi'}$, its expression is

$$\begin{aligned}\alpha^{layer}_{\Xi\to\Xi'} &\equiv \hat{\alpha}^{layer}_{\Xi\to\Xi'}\delta(E_{\Xi} - E_{\Xi'} - E) \text{ with}\\ \hat{\alpha}^{layer}_{\Xi\to\Xi'} &= \frac{\pi e^2\hbar P_0^2}{n_{ref}m_0^2 c\varepsilon_0 E}\left|\left\langle {}^{qn}\zeta_{cb,S} \mid {}^{qn'}\zeta_{hh,X}\right\rangle\right|^2\end{aligned} \tag{5.16}$$

Since all the transitions conserve (k_x, k_y), the transition energy $E_{tr} = E_{\Xi} - E_{\Xi'}$ is given by

$$\begin{aligned}E_{tr} &= E_{z,tr} + \frac{\hbar^2}{2m_{comb}}(k_x^2 + k_y^2) \equiv E_{z,tr} + E_{hor}\\ E_{z,tr} &\equiv E_{z,cb} - E_{z,hh}; \quad \frac{1}{m_{comb}} \equiv \left(\frac{1}{m_{cb}} + \frac{1}{m_{hh}}\right)\end{aligned} \tag{5.17}$$

Note that, for holes, the energies in Eq. (5.4) must be reversed in sign and counted from the VB edge.

To calculate the total absorption coefficient associated to the transitions between two z-bound wavefunctions, the elementary absorption coefficients for all the states per unit of volume must be added, that is $\alpha = (2\pi)^{-2}d_{QW}^{-1}\iint\alpha^{layer}_{\Xi\to\Xi'}\,dk_x dk_y$. $(2\pi)^{-2}$ is the density of states per unit of 4D-volume in the $\{x, y, k_x, k_y\}$ space. Furthermore, there is a density d_{QW}^{-1} of QWs per unit of length, d_{QW} being the period of the QW layers. Thus,

$$\begin{aligned}\alpha &= \int_{E_{z,tr}}^{\infty}\frac{d}{dE}\left(\iint_D\frac{\hat{\alpha}^{layer}_{\Xi\to\Xi'}dk_x dk_y}{(2\pi)^2 d_{QW}}\right)\delta(E - E_{tr})dE \text{ with}\\ D&: k_x^2 + k_y^2 \le 2m_{comb}E_{hor}/\hbar^2\end{aligned} \tag{5.18}$$

where, using Eqs. (3.18), (5.16) and (5.17),

$$\begin{aligned}&\iint_D\frac{\hat{\alpha}^{layer}_{\Xi\to\Xi'}dk_x dk_y}{(2\pi)^2 d_{QW}}\\ &= 2\times\frac{e^2\hbar E_g((m/m_{cb}) - 1)}{8\pi d_{QW}n_{ref}m_0 c\varepsilon_0}\iint_D\frac{\left|\left\langle {}^{qn}\zeta_{cb,S} \mid {}^{qn'}\zeta_{hh,X}\right\rangle\right|^2}{E_{tr}}dk_x dk_y\end{aligned} \tag{5.19}$$

The first factor 2 corresponds to the spin degeneration. The way of solving the D-extended integral as well as Eq. (5.17) is briefly hinted at in the Appendix.

Table 5.2 Properties of QW and barrier materials (calculated as in [12])

Bandgap $GaAs_{0.81}P_{0.19}$ (eV)	1.5414
m_{cb}/m ($GaAs_{0.81}P_{0.19}$)	0.0661346
V_{cb} offset $In_{0.2}Ga_{0.8}As$ (eV)	−0.1647
V_{vb} offset $In_{0.2}Ga_{0.8}As$ (eV)	0.181296
m_{cb}/m_0	0.0556475
m_{hh}/m_0	0.346387
m_{lh}/m_0	0.0774775
QW thickness (nm)	8
Period thickness (nm)	22.6
Total thickness (40 layers, nm)	904

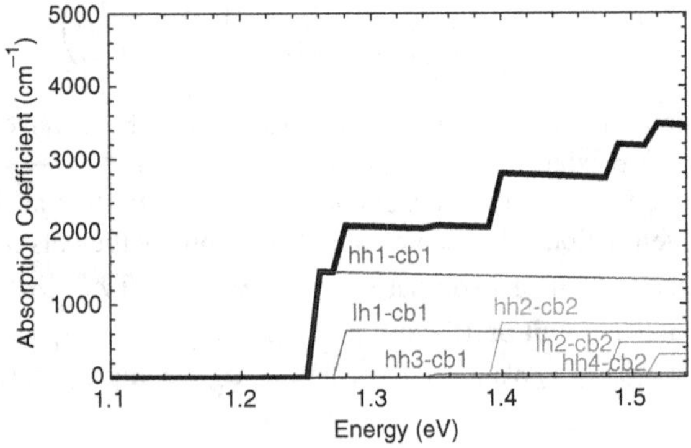

Fig. 5.2 Absorption coefficients for transitions between different sets of states and the total absorption coefficient (*black, thick*) for $\gamma = 0$. The labels show the initial and final band and the initial and final *qn* of the *z*-(bound) component. Reproduced with permission. © 2013. Elsevier [10]

Equation (5.19) is also applicable to vertical unpolarized illumination. As shown in Eq. (2.75), it is the arithmetic mean of the *x* and *y* polarizations, which are equal. For the case of the solar cell described in Tables 5.1 and 5.2, the absorption coefficients between *z*-bound transitions and their sum are presented in Fig. 5.2. For $\gamma = 0$, transitions between states of different parity are forbidden; for $\gamma \neq 0$, they are permitted but negligible.

Under the assumption that all the generation is collected at the electrodes (no recombination), the internal quantum efficiency is calculated as

$$IQE = 1 - \exp(-\alpha W_{QW}) \tag{5.20}$$

(W_{QW} is the total thickness of the QW region) and represented in Fig. 5.3.

In the same figure, the measured external quantum efficiency (EQE) is also presented. Notice that, in the latter case, the current is compared to the incident flux of photons and not to those entering into the semiconductor (the IQE is higher).

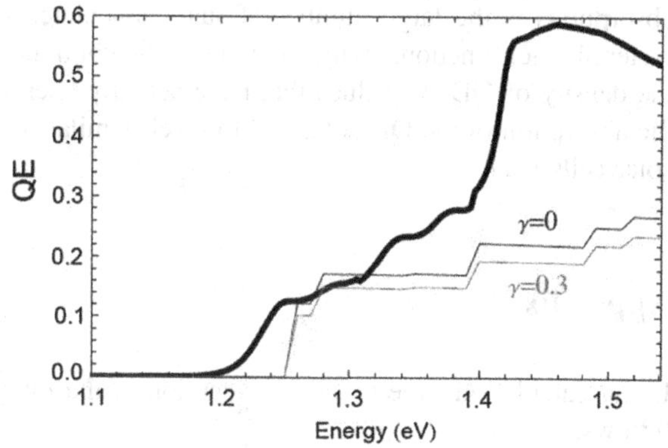

Fig. 5.3 Quantum efficiency (internal for calculations and external for measurements) for the QW solar cell of Table 5.1. Calculations for $\gamma = 0$ and $\gamma = 0.3$ are presented. Reproduced with permission. © 2013. Elsevier [10]

The measured EQE shows a band-to-band generation at 1.42 eV, corresponding to the GaAs emitter, that hides any information about the QW absorption spectrum at higher energies, whereas the calculated value only takes account of the current generated by the QW absorption.

5.4 Concluding Remarks of This Chapter

The EKPH, formerly applied successfully to the study of the envelope functions and the absorption coefficients of quantum dot (QD) solar cells, is now applied to QW solar cells with reasonable semi-quantitative accuracy (the absorption coefficient is approximately reproduced without any fitting).

The measured EQE starts at lower energy, possibly because some of the QWs are wider than planned or their offset is larger than has been calculated [12]. The energy levels of the one-dimensional Hamiltonian appear as well identified corners in the calculated spectrum. Four of these corners appear in the visible part of the IQE (that is, below 1.42 eV) corresponding to the $hh1 \rightarrow cb1$, $lh1 \rightarrow cb1$, $hh3 \rightarrow cb1$ (almost invisible) and $hh2 \rightarrow cb2$ transitions. The same four corners, and perhaps a fifth one, appear (more blurred) in the experimental EQE. Their separation is, however, not the same, probably revealing a potential shape that is not strictly square.

The sub-bandgap photocurrent of QW materials is larger than that of QD materials, but is still small. It is natural to wonder whether this is due to limited absorption or to poor collection of carriers. The results presented in this paper support the explanation of limited absorption. This is interesting because the measured QE may be interpreted as due to photo-generation and therefore be used to optimize solar cell stacks where one of the cells contains QWs.

A radical difference in the matrix element for extended and bound wavefunctions has been stressed. However, this difference is only part of the explanation of why the QDs absorb much less than the QWs. The main reason for the different absorptions is the large number of states associated to the (x, y) part of the QW material eigenfunctions, which belong to the continuous spectrum. The increase of the density of QDs to values that have actually been achieved [20] could increase the absorption of the QD solar cell to levels similar or even higher than that of QW solar cells [21].

Appendix

In our calculations, the practical execution of the integral of Eq. (5.19) is made as follows:

The function inside the integral depends on the array of horizontal wavevectors (k_x, k_y). We build an array containing in each term $\left\{ E_{hor}, \frac{\left|\left\langle {}^{qn}\zeta_{cb,S} \middle| {}^{qn'}\zeta_{hh,X} \right\rangle\right|^2}{E_{tr}} \right\}$, both elements being functions of the horizontal wavevector. The integral, restricted to a domain D, must be a function of the energy so that we must sum only those terms of $\frac{\left|\left\langle {}^{qn}\zeta_{cb,S} \middle| {}^{qn'}\zeta_{hh,X} \right\rangle\right|^2}{E_{tr}}$ such that their corresponding E_{hor} is smaller than a given value of E_{hor}, building in this way the function of E_{hor}. This function, which rather smooth, is interpolated by a polynomial and its derivative is then taken for use in Eq. (5.18).

References

1. Levine BF (1993) Quantum-well infrared photodetectors. J Appl Phys 74(8):R1–R81
2. Barnham KWJ, Duggan G (1990) A new approach to high-efficiency multi-band-gap solar-cells. J Appl Phys 67(7):3490–3493
3. Ekins-Daukes NJ, Barnham KWJ, Connolly JP, Roberts JS, Clark JC, Hill G, Mazzer M (1999) Strain-balanced GaAsP/InGaAs quantum well solar cells. Appl Phys Lett 75(26):4195–4197
4. Okada Y, Shiotsuka N, Takeda T (2004) Potentially modulated multi-quantum wells for high-efficiency solar cell applications. Sol Energy Mater Sol Cells 85(2):143–152
5. Kailuweit P, Kellenbenz R, Philipps SP, Guter W, Bett AW, Dimroth F (2010) Numerical simulation and modeling of GaAs quantum-well solar cells. J Appl Phys 107(6):064317. doi:10.1063/1.3354055
6. Luque A, Marti A, Antolín E, Linares PG, Tobías I, Ramiro I, Hernandez E (2011) New Hamiltonian for a better understanding of the Quantum Dot Intermediate Band Solar Cells. Sol Energy Mater Solar Cells 95:2095–2101. doi:10.1016/j.solmat.2011.02.028
7. Luque A, Mellor A, Antolin E, Linares PG, Ramiro I, Tobias I, Marti A (2012) Symmetry considerations in the empirical k.p Hamiltonian for the study of intermediate band solar cells. Sol Energy Mater Sol Cells 103:171–183
8. Martí A, Cuadra L, Luque A (2000) Quantum dot intermediate band solar cell. Proceedings of the 28th IEEE photovoltaics specialists conference, pp 940–943. New York
9. Luque A, Martí A (1997) Increasing the efficiency of ideal solar cells by photon induced transitions at intermediate levels. Phys Rev Lett 78(26):5014–5017

10. Luque A, Antolín E, Linares PG, Ramiro I, Mellor A, Tobías I, Martí A (2013) Interband optical absorption in quantum well solar cells. Solar Energy Mater Solar Cells 112:20–26. doi:10.1016/j.solmat.2012.12.045
11. Luque A, Marti A, Mellor A, Marron DF, Tobias I, Antolín E (2012) Absorption coefficient for the intraband transitions in quantum dot materials. Prog Photovoltaics Res Appl. doi:10.1002/pip.1250
12. Linares PG, Marti A, Antolin E, Luque A (2011) III-V compound semiconductor screening for implementing quantum dot intermediate band solar cells. J Appl Phys 109:014313
13. Messiah A (1960) Mécanique Quantique. Dunod, Paris
14. Luque A, Marti A, Mendes MJ, Tobias I (2008) Light absorption in the near field around surface plasmon polaritons. J Appl Phys 104(11):113118. doi:10.1063/1.3014035
15. Harrison P (2000) Quantum wells wires and dots. Wiley, New York
16. Coon DD, Karunasiri RPG (1984) New mode of IR detection using quantum wells. Appl Phys Lett 45(6):649–651. doi:10.1063/1.95343
17. Datta S (1989) Quantum phenomena. Molecular Series on Solid State Devices, vol 8. Addison Wesley, Reading (Mass)
18. Luque A, Marti A, Antolín E, Linares PG, Tobias I, Ramiro I (2011) Radiative thermal escape in intermediate band solar cells. AIP Adv 1:022125
19. Fermi E (1932) Quantum theory of radiation. Rev Mod Phys 4(1):0087–0132
20. Akahane K, Kawamura T, Okino K, Koyama H, Lan S, Okada Y, Kawabe M, Tosa M (1998) Highly packed InGaAs quantum dots on GaAs(311)B. Appl Phys Lett 73(23):3411–3413
21. Tobías I, Luque A, Antolín E, Linares PG, Ramiro I, Hernández E, Martí A (2012) Realistic performance prediction in nanostructured solar cells. J Appl Phys 112:24518. doi:10.1063/1.4770464

Chapter 6
Interband Optical Absorption in Homogeneous Semiconductors

Abstract In this chapter, the Empiric k·p Hamiltonian method is applied to the case of semiconductors that are not nanostructured. In this case, the solutions of the diagonalized Hamiltonian are plane waves. The general procedure is applied to obtain the envelope functions. These are plane waves and the calculation of the elements of matrix is very simple. Perhaps the biggest complexity is in the integration of the huge number of transitions involved. The procedure is applied to GaAs and the interband absorption coefficients are calculated. They result in reasonable agreement with the measured values, proving again the appropriateness of the Empiric k·p Hamiltonian method.

Keywords Quantum calculations · k·p methods · GaAs absorption

Calculations of band-to-band transitions are one of the first achievements of the solid state theory. The topic is widely and successfully studied by many authors. The purpose of this chapter is to show that the EHPH may also be used for such calculations to within a reasonable accuracy. This chapter provides the only example in this book of the application of the EKPH method to a homogeneous semiconductor without any nanostructure. It follows closely the steps of reference [1].

6.1 Fourier Transform Reminder

As stated in Chap. 3, Sect. 3.1.4, the use of some kind of Fourier transform is a practical way of obtaining a plane wave development. In some cases, the wavefunctions $\Phi(\boldsymbol{r}) = \xi(x)\varphi(y)\zeta(z)$ are the product of three one-dimensional functions. In this case, the Fourier transform is the product of three one dimensional Fourier transforms. In general, the three-dimensional Fourier transform is obtained by the successive application of three one-dimensional transforms. In this book, the Fourier Integrals (FIs) are used for extended wavefunctions.

A. Luque and A.V. Mellor, *Photon Absorption Models in Nanostructured Semiconductor Solar Cells and Devices*, SpringerBriefs in Applied Sciences and Technology, DOI 10.1007/978-3-319-14538-9_6

$$\begin{aligned} \sqrt{\frac{L_x}{2\pi}}\phi_x &= \mathscr{F}[\xi(x)](k_x) \equiv \frac{1}{\sqrt{2\pi}} \int_{-\infty}^{\infty} \xi(x)e^{-ik_x x}dx \\ \sqrt{\frac{2\pi}{L_x}}\xi(x) &= \mathscr{F}^{-1}[\phi(k_x)](x) = \frac{1}{\sqrt{2\pi}} \int_{-\infty}^{\infty} \phi(k_x)e^{ik_x x}dk_x \end{aligned} \tag{6.1}$$

The Fourier transform is denoted with a script F and its argument is between square brackets. This argument is a function that shows explicitly its variable for the FI and its inverse (IFI); the square brackets are followed by the transform variable between parentheses (or round brackets).

6.2 Envelope Functions

The calculation of the envelope functions follows the subsequent path [2, 3]: (1) resolution of the diagonalized TISE; (2) calculation of the solution's Fourier transform; (3) multiplication by the appropriate element of the transformation matrix; (4) calculation of the inverse Fourier transform.

6.2.1 Solutions of the Time Independent Schrödinger Equation for the Diagonalized Hamiltonian

For a homogeneous semiconductor the diagonalized TISE is,

$$-\frac{\hbar^2}{2m^*}\nabla^2\Phi(\boldsymbol{r}) = E\Phi(\boldsymbol{r}) \tag{6.2}$$

The asterisk refers to the effective mass in the band we are considering: the conduction band (cb) or the light hole (lh), heavy hole (hh) and split-off (so) VBs. We can separate variables and write the solution as

$$\Phi(\boldsymbol{r}) = \exp(k'_x x + k'_y y + k'_z z)\Big/\sqrt{L_x L_y L_z} = \exp(\boldsymbol{k}'\cdot\boldsymbol{r})\Big/\sqrt{\Omega} \tag{6.3}$$

The eigenenergy of an eigenfunction is

$$E = \frac{\hbar^2 k_x'^2}{2m^*} + \frac{\hbar^2 k_y'^2}{2m^*} + \frac{\hbar^2 k_z'^2}{2m^*} = \frac{\hbar^2 k'^2}{2m^*} \tag{6.4}$$

6.2.2 Calculating Fourier Transforms

The plane-wave development coefficients of $\Phi(\boldsymbol{r})$ in Eq. (6.3) can be directly written in terms of Fourier transforms. In this case, all the functions are extended plane waves; we recall that $\mathcal{F}\left[e^{ik'_x x}\right](k_x) = \sqrt{2\pi}\delta(k'_x - k_x)$, therefore,

$$\begin{aligned}\phi(\boldsymbol{k}) &= \phi_z(k_z)\phi_x(k_x)\phi_y(k_y)\\ &= \frac{(2\pi)^{3/2}}{\Omega^{1/2}}\mathcal{F}\left[\frac{e^{ik'_x x}}{\sqrt{L_x}}\right](k_x)\mathcal{F}\left[\frac{e^{ik'_y y}}{\sqrt{L_y}}\right](k_y)\mathcal{F}\left[\frac{e^{ik'_z z}}{\sqrt{L_z}}\right](k_z)\\ &= \frac{(2\pi)^3}{\Omega}\delta(k'_x - k_x)\delta(k'_y - k_y)\delta(k'_z - k_z)\end{aligned} \tag{6.5}$$

where all normalizing and conversion factors in Sect. 3.1.4 have been considered [4].

6.2.3 Calculating the Envelope Functions

To calculate the envelope functions, $\phi(\boldsymbol{k})$ is to be multiplied by one of the elements ${}^{\gamma}T_{i,j}(\boldsymbol{k})$. The row of the element, from 1 to 4, corresponds to the *cb*, *lh*, *hh* and *so* bands respectively. The column of the element, from 1 to 4 corresponds to the *X*-, *Y*-, *Z*- and *S*-envelopes respectively [3]. The IFI is calculated with respect to the variables *x*, *y* and *z*. We obtain

$$\begin{aligned}\Psi(r) &= \xi(x)\psi(y)\zeta(z) = \frac{(\Omega)^{1/2}}{(2\pi)^{3/2}}\mathcal{F}^{-1}\left[{}^{\gamma}T_{i,j}(k_x,k_y,k_z)\phi(k_x,k_y,k_z)\right](x,y,z)\\ &= \frac{(2\pi)^{3/2}}{\Omega^{1/2}}\mathcal{F}^{-1}\left[{}^{\gamma}T_{i,j}(k_x,k_y,k_z)\delta(k'_x-k_x)\delta(k'_y-k_y)\delta(k'_z-k_z)\right](x,y,z)\\ &= \frac{1}{\Omega^{1/2}}{}^{\gamma}T_{i,j}(k'_x,k'_y,k'_z)\mathcal{F}^{-1}\left[(2\pi)^{3/2}\delta(k'_x-k_x)\delta(k'_y-k_y)\delta(k'_z-k_z)\right](x,y,z)\\ &= T_{i,j}(k'_x,k'_y,k'_z)\frac{e^{i(k'_x x+k'_y y+k'_z z)}}{\Omega^{1/2}}\end{aligned} \tag{6.6}$$

Depending on the T selected, it will correspond to the projection onto a GBF and will belong to a certain band. In summary, all the envelopes are plane waves with the same wavevector: that of the diagonalized EKPH. The wavevectors are projected onto the bands according to the corresponding element of the matrix (T).

6.3 Photon Absorption

6.3.1 Dipole Matrix Elements

As discussed in Chap. 5, the dipole matrix element ruling the photon-electron interaction is $\langle \Xi | \boldsymbol{\epsilon} \cdot \boldsymbol{p} | \Xi \rangle$ [5, 6] where $\boldsymbol{\epsilon}$ is the light-polarization vector and $\boldsymbol{p}$ the momentum operator (proportional to the gradient). Let us consider Ψ to be slowly varying plane waves; this is true for small wavevectors. $\boldsymbol{p}(u\Psi) = u\boldsymbol{p}\Psi + \Psi\boldsymbol{p}u$. Due to the different scale of distances associated to the k·p method, the integral factorization rule is considered applicable and $\langle u\Psi | \boldsymbol{\epsilon} \cdot \boldsymbol{p} | u'\Psi' \rangle \cong \langle u | u' \rangle \langle \Psi | \boldsymbol{\epsilon} \cdot \boldsymbol{p} | \Psi \rangle + \langle u | \boldsymbol{\epsilon} \cdot \boldsymbol{p} | u' \rangle \langle \Psi | \Psi' \rangle$ [7–9], where the u functions are normalized according to the cell-averaged rule discussed in Sect. 2.1.2. Thus

$$
\begin{aligned}
\langle \Xi | \epsilon \cdot p | \Xi' \rangle &= \sum_{v,v'} \langle u_{0,v} | u_{0,v'} \rangle \langle \Psi_v | \epsilon \cdot p | \Psi'_{v'} \rangle + \sum_{v,v'} \langle u_{0,v} | \epsilon \cdot p | u_{0,v'} \rangle \langle \Psi_u | \Psi'_{u'} \rangle \\
&= \sum_{v} \langle \Psi_v | \epsilon \cdot p | \Psi'_{v'} \rangle + \sum_{v,v'} \langle u_{0,v} | \epsilon \cdot p | u_{0,v'} \rangle \langle \Psi_u | \Psi'_{u'} \rangle
\end{aligned}
\tag{6.7}
$$

where we have also considered that $\langle u_{0v} | u_{0v'} \rangle = \boldsymbol{\delta}_{v,v'}$.

Any polarization orientation may be developed into its coordinate axes [3, 10] leading to terms containing p_x, p_y, and p_z. For the calculation of $\langle u_{0v} | \boldsymbol{\epsilon} \cdot \boldsymbol{p} | u_{0v'} \rangle$, we can use the symmetry properties of the us in zincblende crystals. The only non-null terms [2, 9] are

$$
\begin{aligned}
&\langle S | p_x | X \rangle = \langle S | p_y | Y \rangle = \langle S | p_z | Z \rangle = P_0 \\
&\text{with} \quad P_0 = \sqrt{\frac{E_g}{2} m_0 \left(\frac{m_0}{m^*_{cb}} - 1 \right)}
\end{aligned}
\tag{6.8}
$$

E_g being the (barrier) semiconductor's bandgap, m_0 the mass of the electron in the vacuum and m^*_{cb} its effective mass in the conduction band. Thus, for the case of x-polarization, Eq. (6.7) becomes

$$
\begin{aligned}
\langle \Xi | p_x | \Xi' \rangle &= \sum_{v} \langle \Psi_v | p_x | \Psi'_v \rangle + P_0 \langle \Psi_S \mid \Psi'_X \rangle \equiv \Pi_x + P_0 \langle \Psi_S \mid \Psi'_X \rangle \\
\Pi_x &\equiv \langle \Psi_X | p_x | \Psi'_X \rangle + \langle \Psi_Y | p_x | \Psi'_Y \rangle + \langle \Psi_Z | p_x | \Psi'_Z \rangle + \langle \Psi_S | p_x | \Psi'_S \rangle \\
&\equiv \Pi_{xX} + \Pi_{xY} + \Pi_{xZ} + \Pi_{xS}
\end{aligned}
\tag{6.9}
$$

Vertical illumination involves photons with horizontal polarization. Let us examine, for example, the Π_{xS} matrix element for a $hh \rightarrow cb$ transition. It is a function of the differences $k_{x,hh} - k_{x,cv} = \omega_x$, $k_{y,hh} - k_{y,cv} = \omega_y$ and $k_{z,hh} - k_{z,cv} = \omega_z$. Thus,

$$\Pi_{xS} = T^*_{cb,S}(\boldsymbol{k}_{cv})T_{hh,S}(\boldsymbol{k}_{hh}) \int_{-L_z/2}^{L_z/2} \frac{e^{i\omega_z z}}{L_z} dy \int_{-L_y/2}^{L_y/2} \frac{e^{i\omega_y y}}{L_y} dy \int_{-L_x/2}^{L_x/2} \frac{e^{-ik_{cb,x}x} p_x e^{ik_{hh,x}x}}{L_x} dx \tag{6.10}$$

The only permitted k components are integer multiples of $2\pi/L$, and the same for their differences ω. The y- and z-functions are one when $\omega_y \to 0$ and are strictly zero for the permitted lateral differences, of $\pm 2\pi/L, \pm 4\pi/L$, etc. Thus, k_y and k_z are conserved. The x-function is $\hbar k_{hh,x}$ for $\omega_x \to 0$ and zero for the permitted lateral values. Thus, k_x is also conserved. In summary, by dropping the band index from the k-components,

$$\Pi_{xS} = \hbar k_x T^*_{cb,S}(\boldsymbol{k})T_{hh,S}(\boldsymbol{k}) \tag{6.11}$$

The same is applicable to other matrix elements. Therefore,

$$\Pi_x = \hbar k_x \left(T^*_{cb,X}(\boldsymbol{k})T_{hh,X}(\boldsymbol{k}) + T^*_{cb,Y}(\boldsymbol{k})T_{hh,Y}(\boldsymbol{k}) + T^*_{cb,Z}(\boldsymbol{k})T_{hh,Z}(\boldsymbol{k}) + T^*_{cb,S}(\boldsymbol{k})T_{hh,S}(\boldsymbol{k}) \right) = 0 \tag{6.12}$$

(the envelope internal product without envelope subindex means, by definition, the sum of internal products). It is zero because the T matrix is hermitic.

Thus, Eq. (6.9) can be rewritten as

$$\left\langle {}^{qn}\Xi_{cb} | p_x | {}^{qn'}\Xi_{hh} \right\rangle = P_0 \left\langle \Psi_{cb,S} \mid \Psi_{hh,X} \right\rangle = P_0 T^*_{cb,S}(\boldsymbol{k})T_{hh,X}(\boldsymbol{k}) \tag{6.13}$$

A similar treatment is to be used for y- and z- polarization. The z-polarization, not possible if the illumination is purely vertical. Note that $\boldsymbol{k}$ refers now to the selected wavevector, affected of a prime in the preceding section to distiguish it from the Fourier transform variable.

6.3.2 The Absorption Coefficient for x- or y-Polarization

The number of photon absorptions per unit of time [2, 5] when the electron passes from state $|\Xi\rangle$ to state $|\Xi'\rangle$ is

$$w_{\Xi \to \Xi'} = \Delta n_{ph} \frac{\pi e^2 \hbar}{n_{ref}^2 m^2 \varepsilon_0 E} |\langle \Xi | \boldsymbol{p} \cdot \boldsymbol{\epsilon} | \Xi' \rangle|^2 \delta(E_{\Xi} - E_{\Xi'} - E) \tag{6.14}$$

where ε_0 is the vacuum permittivity, n_{ref} is the refractive index of the medium and Δn_{ph} is the density of photons in all the modes of energy E. This density can be related to the photon flux by $\Delta F_{ph} = (c/n_{ref})\Delta n_{ph}$. Taking into account that the number of photon absorptions per unit of time is related to the elementary absorption coefficient $\alpha_{\Xi\to\Xi'}$ in the transition under consideration by $w_{\Xi\to\Xi'} = \Delta F_{ph}\alpha_{\Xi\to\Xi'}$, its expression is

$$\begin{aligned} \alpha_{\Xi\to\Xi'} &= \alpha'_{\Xi\to\Xi'}\delta(E_\Xi - E_{\Xi'} - E) \\ \text{with } \alpha'_{\Xi\to\Xi'} &= \frac{\pi e^2\hbar P_0^2}{n_{ref}m^2 c\varepsilon_0 E}\left|T^*_{cb,S}(k)T_{hh,X}(\boldsymbol{k})\right|^2 \end{aligned} \tag{6.15}$$

Since all the transitions conserve ($\boldsymbol{k}$), the transition energy $E_{tr} = E_\Xi - E_{\Xi'}$ is given by

$$\begin{aligned} E_{tr} &= E_g + \frac{\hbar^2}{2m^*_{comb}}(k_x^2 + k_y^2 + k_2^z) = E_g + E_{kin} \\ E_{tr} &\equiv E_{cb} - E_{hh}; \quad \frac{1}{m^*_{comb}} \equiv \left(\frac{1}{m^*_{cb}} + \frac{1}{m^*_{hh}}\right) \end{aligned} \tag{6.16}$$

Note that, for holes, the energies in Eq. (6.4) must be reversed in sign and counted from the VB edge.

To calculate the total absorption coefficient, the elementary absorption coefficients for all the states per unit of volume must be added, that is $\alpha = (2\pi)^{-3}\iiint \alpha_{\Xi\to\Xi'}dk_x dk_y dk_z$. $(2\pi)^{-3}$ is the density of states per unit of 6D-volume in the $\{x, y, z, k_x, k_y, k_z\}$ space. Thus,

$$\begin{aligned} \alpha &= \int_{Eg}^{\infty} \frac{d}{dE}\left(\iiint_D \frac{\alpha' dk_x dk_y dk_y}{(2\pi)^3}\right)\delta(E - E_{tr})dE \\ &\text{with } D:\ k_x^2 + k_y^2 + k_y^2 \le 2m^*_{comb}E_{kin}/\hbar^2 \end{aligned} \tag{6.17}$$

where, using (6.8), (6.15) and (6.17),

$$\begin{aligned} &\iiint_D \frac{\alpha' dk_x dk_y dk_y}{(2\pi)^2} \\ &= 2\times\frac{e^2\hbar E_g((m/m^*_{cb}) - 1)}{16\pi^2 n_{ref}mc\varepsilon_0}\iiint_{D(E_{kin})}\frac{\left|T^*_{cb,S}(\boldsymbol{k})T_{hh,X}(\boldsymbol{k})\right|^2}{E_{kin} + E_g}dk_x dk_y dk_y \end{aligned} \tag{6.18}$$

The first first factor 2 corresponds to the spin degeneration. The procedure for obtaining this integral is based on the Appendix of Chap. 5.

We present in Fig. 6.1 the results of these equations together with the a classical measurement [11] similar results may be found in other classical works [12, 13]. The calculations have been made with the data of Table 6.1.

The curve to compare is the one labeled “High purity”. We can observe a reasonable agreement that obtained in some few minutes with a laptop. This

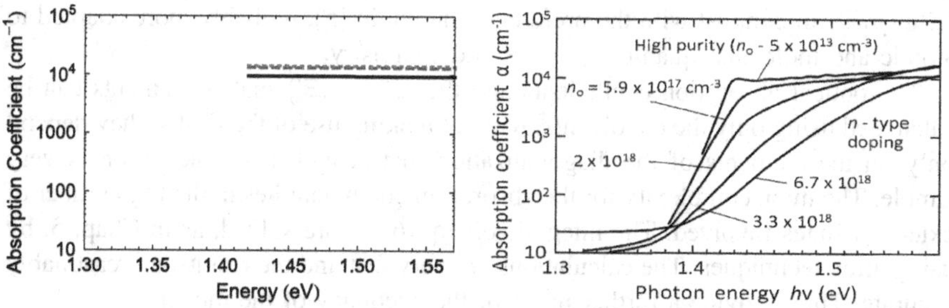

Fig. 6.1 *Left*, calculated absorption coefficient for band to band transitions in GaAs for $\gamma = 0$ (*red, dot-dashed line*), 2 (*blue, dashed line*) and 5 (*black, solid line*). Reproduced with permission, © 2014, Elsevier [1]. *Right*, measured absorption coefficient after [11]. Reproduced with permission, © 1975, AIP

Table 6.1 GaAs parameters for calculations

Bandgap GaAs (300 K)	1.42
m_{cb}/m_0	0.0613
m_{hh}/m_0	0.35
m_{lh}/m_0	0.09

supports the adequacy of the use of the EKPH model although we do not pretend to replace this method by the established ones in this widely studied topic.

The GaAs absorption coefficient takes values much higher, approaching to or exceeding coefficients 10^6 cm^{-1} for bigger energies (e.g. 4 eV). This is due to the enabling of transitions to between more couples of bands. The subject is beyond of the scope of this book. Our calculations are limited to the transitions between VB and CB.

6.4 Concluding Remarks of This Chapter

The EKPH model has been applied to the calculation of the eigenfunctions and the band-to-band absorption of the GaAs. For it the envelope functions are calculated: they are plane waves. In consequence, in this model, the wavefunctions are linear combinations of plane waves multiplied by the Γ point Bloch function (the GBFs) for the four bands involved: *cb*, *lh*, *hh* and *so*. In this respect they are close to the classical Bloch function, in which the periodic part is dependent on the wavevector (and difficult to know) but they are not mixed with the other bands. The module of the envelopes is given by the diagonalization element of matrix corresponding to the band of the eigenvector and to the concerned GBF.

In this chapter the optical elements of matrix are of the form $\langle \Xi | \boldsymbol{\epsilon} \cdot \boldsymbol{p} | \Xi' \rangle$ and not $\langle \Xi | \boldsymbol{\epsilon} \cdot \boldsymbol{r} | \Xi' \rangle$ as in Chaps. 1–4. This is obliged by the extended nature of all the functions. This also explain the preference of most solid state physics for the

$\langle \Xi | \boldsymbol{\epsilon} \cdot \boldsymbol{p} | \Xi' \rangle$ in contrast with the preference shown in [5], probably more adapted to atomic and molecular quantum physics and chemistry.

The optical absorption coefficients and the $\langle \Xi | \boldsymbol{\epsilon} \cdot \boldsymbol{p} | \Xi' \rangle$ matrix elements can be calculated using only the envolvents, without making use of the GBFs; they depend only on the elements of the diagonalization matrix and their calculation is very simple. The main complexity for the absorption coefficient lies in the big number of extended states involved. The integrals of Eq. (6.18) are solved, as in Chap. 5, by using array techniques. The calculations are very fast and the results are reasonably accurate. They provide a further proof of the adequacy of the model.

References

1. Luque A, Panchak A, Mellor A, Vlasov A, Martí A, Andreev V (2014) Empiric k·p Hamiltonian calculation of the band-to-band photon absorption in semiconductors. Physica B. doi:http://dx.doi.org/10.1016/j.physb.2014.08.026
2. Luque A, Marti A, Antolín E, Linares PG, Tobías I, Ramiro I, Hernandez E (2011) New Hamiltonian for a better understanding of the quantum dot intermediate band solar cells. Solar Energy Mater Solar Cells 95:2095–2101
3. Luque A, Mellor A, Antolin E, Linares PG, Ramiro I, Tobias I, Marti A (2012) Symmetry considerations in the empirical k.p Hamiltonian for the study of intermediate band solar cells. Solar Energy Mater Sol Cells 103:171–183
4. Luque A, Antolín E, Linares PG, Ramiro I, Mellor A, Tobías I, Martí A (2013) Interband optical absorption in quantum well solar cells. Solar Energy Mater Solar Cells 112:20–26. doi:10.1016/j.solmat.2012.12.045
5. Messiah A (1960) Mécanique quantique. Dunod, Paris
6. Luque A, Marti A, Mendes MJ, Tobias I (2008) Light absorption in the near field around surface plasmon polaritons. J Appl Phys 104(11):113118. doi:10.1063/1.3014035
7. Harrison P (2000) Quantum wells wires and dots. Wiley, New York
8. Coon DD, Karunasiri RPG (1984) New mode of IR detection using quantum wells. Appl Phys Lett 45(6):649–651. doi:10.1063/1.95343
9. Datta S (1989) Quantum phenomena. Molecular Series on Solid State Devices, vol 8. Addison Wesley, Reading (Mass)
10. Luque A, Marti A, Antolín E, Linares PG, Tobias I, Ramiro I (2011) Radiative thermal escape in intermediate band solar cells. AIP Adv 1:022125
11. Casey HC, Sell DD, Wecht KW (1975) Concentration-dependence of absorption-coefficient for n-type and p-type GaAs between 1.3 and 1.6 eV. J Appl Phys 46(1):250–257. doi:10.1063/1.321330
12. Hobden MV, Sturge MD (1961) Optical absorption edge of gallium arsenide. Proc Phys Soc London 78(502):615. doi:10.1088/0370-1328/78/4/119
13. Sturge MD (1962) Optical absorption of gallium arsenide between 0.6 and 2.75 eV Physical Rev 127(3):768. doi:10.1103/PhysRev.127.768

Chapter 7
Comparing the Eight-Band Luttinger-Kohn-Pikus-Bir-Hamiltonian with the Four-Band Empiric k·p Hamiltonian

Abstract The purpose of this chapter is to compare on one side the Luttinger Kohn Hamiltonian with the Pikus Bir modification to account for the strained lattices of nanostructured semiconductors, commonly used by the solid state physicists, with, on the other side, the Empiric k·p Hamiltonian proposed and described in Chap. 3, and more oriented to the research and development of optoelectronic devices. The concept of spin is explained, mainly to fix the nomenclature. It is pointed out that the $|S\rangle, |X\rangle, |Y\rangle, |Z\rangle$ Γ-Bloch functions used in Chap. 3 may have up and down spins becoming $|S\uparrow\rangle, |S\downarrow\rangle\ldots$ etc. However these are not the Γ-Bloch functions in this chapter. These are instead $|cb+\rangle, |hh+\rangle, |lh+\rangle, |so+\rangle$, and their four counterparts with minus signs, which are linear combinations of $|S\uparrow\rangle, |S\downarrow\rangle\ldots$ etc. The eight-band Luttinger Kohn Pikus Bir Hamiltonian is described, making it possible for the reader to use it with a strain distribution. An approximate four-band Luttinger Kohn Pikus Bir Hamiltonian is also presented. Fittings are made to match the calculated values with the experimental values of the effective masses and bandgaps. The eigenvalues and eigenfunctions of the new Hamiltonians are calculated, and subsequently the sub-bandgap absorption coefficients and the quantum efficiency for the exemplary InAs/GaAs cell modeled along this book. The latter is compared with the experimental data and with the calculations based on the Empiric k·p Hamiltonian. An assessment of the calculation time with the different methods is presented.

Keywords Quantum calculations · Eight-band k·p Hamiltonian · Four-band k·p Hamiltonians · Energy spectra · Photon absorption calculations

7.1 Spinors

A consequence of relativity in quantum mechanics is the electron spin [1, 2]. Because of it, the states may now be described as spinors. A spinor is an algebraic entity, which, like a vector, has several components. In the case of the electron, two components are required, and the spinor is represented as

A. Luque and A.V. Mellor, *Photon Absorption Models in Nanostructured Semiconductor Solar Cells and Devices*, SpringerBriefs in Applied Sciences and Technology, DOI 10.1007/978-3-319-14538-9_7

$$|g\rangle = \begin{pmatrix} \alpha(\boldsymbol{r}) \\ \beta(\boldsymbol{r}) \end{pmatrix} \tag{7.1}$$

Any operator, like the Hamiltonian is represented by a matrix of the type

$$\begin{pmatrix} H & 0 \\ 0 & H \end{pmatrix} \tag{7.2}$$

acting on the spinor of Eq. (7.1). The off-diagonal terms may also be non-zero, as will be seen later.

The internal product of two spinors is

$$\langle g' \mid g\rangle = \int \alpha'^{*} \alpha d^3 r + \int \beta'^{*} \beta d^3 r \tag{7.3}$$

As studied so far, the Bloch functions at the Γ point, denoted GBFs, were functions without spin. In reality it meant that they were of the type

$$\begin{pmatrix} |S\rangle \\ 0 \end{pmatrix} \quad \text{and} \quad \begin{pmatrix} 0 \\ |S\rangle \end{pmatrix}$$

or, with a more compact notation, $|S\uparrow\rangle$ and $|S\downarrow\rangle$ respectively (called S up and S down or S plus and S minus). The same is true for the rest of the GBFs. It is clear that the GBFs of different spin are orthogonal. In general, with this notation, the generic state is the sum of a spin up state and a spin down state.

$$|g\rangle = |\alpha\uparrow\rangle + |\beta\downarrow\rangle \tag{7.4}$$

7.2 Spin Orbit Coupling in Gamma-Point Bloch Functions

The spin acquires energetic importance in the presence of a magnetic field $\boldsymbol{B}$. The Hamiltonian correction due to it is

$$H_B = \mu_B(\sigma_x B_x + \sigma_y B_y + \sigma_z B_z) \equiv \mu_B \boldsymbol{\sigma} \cdot \boldsymbol{B} \tag{7.5}$$

where μ_B is the Bohr magnetron $e\hbar/2m_0$ (in the SI) and

$$\sigma_x = \begin{pmatrix} 0 & 1 \\ 1 & 0 \end{pmatrix}; \quad \sigma_y = \begin{pmatrix} 0 & -i \\ i & 0 \end{pmatrix}; \quad \sigma_z = \begin{pmatrix} 1 & 0 \\ 0 & -1 \end{pmatrix}; \tag{7.6}$$

are the Pauli matrices. A consequence is that when a magnetic field is present, the non diagonal terms of the Hamiltonian spinor representation cease to be zero. But even in absence of a magnetic field, strong electric fields are seen as magnetic fields by moving electrons. This is what happens in the atoms and the spin-orbit coupling term is

$$(H_{so}) = \frac{-i\hbar^2}{4m_0^2c^2}\boldsymbol{\sigma}\cdot(\nabla U_L \times \nabla) \tag{7.7}$$

where (H_{so}) is a spinor-space matrix operator whose diagonal terms are zero. In the following, the nomenclature $(H_{so})_{\uparrow\downarrow}$ represents the $(H_{so})_{1,2}$ two-dimensional matrix element in spinors space, and so on. We also follow the nomenclature of Chap. 2 so that U_L is the lattice potential. In zincblende semiconductors, the non-zero elements of the k·p matrix development are

$$\begin{aligned}\langle X\uparrow|(H_{so})_{\uparrow\downarrow}|Z\downarrow\rangle &= -\langle X\uparrow|(H_{so})_{\downarrow\uparrow}|Z\downarrow\rangle\\ &= i\langle Y\uparrow|(H_{so})_{\uparrow\downarrow}|Z\downarrow\rangle\\ &= i\langle Y\uparrow|(H_{so})_{\downarrow\uparrow}|Z\downarrow\rangle\\ &= i\langle X\uparrow|(H_{so})_{\uparrow\uparrow}|Y\uparrow\rangle\\ &= -i\langle X\downarrow|(H_{so})_{\downarrow\downarrow}|Y\downarrow\rangle\\ &\equiv \Delta/3\end{aligned} \tag{7.8}$$

along with their complex conjugates, which result from interchanging bras and kets. Thus, the single real parameter Δ, in general experimentally obtained, may characterize the spin orbit coupling.

7.2.1 Gamma-Point Eigenvalues and Eigenfunctions

Equation (7.8) must be used to write the matrix $\langle 0, \nu' \updownarrow|(H) + (H_{so})|0, \nu' \updownarrow\rangle$ in Fig. 7.1 (the double pointed arrows means that the GBF may be spin up or down).

Note that the elements of the basis have been ordered so that diagonal blocks are formed. This simplifies the search for the eigenvalues and eigenvectors; only two

$\langle 0,\nu'\updownarrow\|(H)+(H_{so})\|0,\nu'\updownarrow\rangle$	$\|S\uparrow\rangle$	$\|S\downarrow\rangle$	$\|X\uparrow\rangle$	$\|Y\uparrow\rangle$	$\|Z\downarrow\rangle$	$\|X\downarrow\rangle$	$\|Y\downarrow\rangle$	$\|Z\uparrow\rangle$
$\langle S\uparrow\|$	E_{CB}	0	0	0	0	0	0	0
$\langle S\downarrow\|$	0	E_{CB}	0	0	0	0	0	0
$\langle X\uparrow\|$	0	0	E'_{VB}	$-i\Delta/3$	$\Delta/3$	0	0	0
$\langle Y\uparrow\|$	0	0	$i\Delta/3$	E'_{VB}	$-i\Delta/3$	0	0	0
$\langle Z\downarrow\|$	0	0	$\Delta/3$	$i\Delta/3$	E'_{VB}	0	0	0
$\langle X\downarrow\|$	0	0	0	0	0	E'_{VB}	$i\Delta/3$	$-\Delta/3$
$\langle Y\downarrow\|$	0	0	0	0	0	$-i\Delta/3$	E'_{VB}	$-i\Delta/3$
$\langle Z\uparrow\|$	0	0	0	0	0	$-\Delta/3$	$i\Delta/3$	E'_{VB}

Fig. 7.1 Hamiltonian matrix of the Γ-point Bloch functions taking into account the spin-orbit interaction

and three dimensional matrices have to be diagonalized. The eigenvalues and eigenvectors (eigenstates) are,

$$
\begin{array}{ll}
\text{Eigenvalue} & \text{Eigenstate} \\
E_{CB} & |cb+\rangle = |S\uparrow\rangle \\
E_{CB} & |cb-\rangle = |S\downarrow\rangle \\
E'_{VB} + \Delta/3 & |hh+\rangle = \frac{1}{\sqrt{2}}|X\uparrow\rangle + \frac{i}{\sqrt{2}}|Y\uparrow\rangle \\
E'_{VB} + \Delta/3 & |hh-\rangle = \frac{i}{\sqrt{2}}|X\downarrow\rangle + \frac{1}{\sqrt{2}}|Y\downarrow\rangle \\
E'_{VB} + \Delta/3 & |lh+\rangle = \frac{i}{\sqrt{6}}|X\downarrow\rangle - \frac{1}{\sqrt{6}}|Y\downarrow\rangle - i\sqrt{\frac{2}{3}}|Z\uparrow\rangle \\
E'_{VB} + \Delta/3 & |lh-\rangle = \frac{1}{\sqrt{6}}|X\uparrow\rangle - \frac{i}{\sqrt{6}}|Y\uparrow\rangle + \sqrt{\frac{2}{3}}|Z\downarrow\rangle \\
E'_{VB} - 2\Delta/3 & |so+\rangle = \frac{1}{\sqrt{3}}|X\uparrow\rangle + \frac{i}{\sqrt{3}}|Y\downarrow\rangle + \frac{1}{\sqrt{3}}|Z\uparrow\rangle \\
E'_{VB} - 2\Delta/3 & |so-\rangle = -\frac{i}{\sqrt{3}}|X\uparrow\rangle - \frac{1}{\sqrt{3}}|Y\uparrow\rangle + \frac{i}{\sqrt{3}}|Z\downarrow\rangle
\end{array}
\tag{7.9}
$$

It can be seen that the conduction band eigenvalue is doubly spin degenerated. The same happens with the heavy and light holes, which are also mutually degenerate. However, due to the spin-orbit coupling, their energy is increased. For this reason we have used a prime in Eq. (7.9) and now set $E_{VB} = E'_{VB} + \Delta/3$. The split off band energy is below that of the heavy and light holes, as is the case in Chap. 3.

We have given a new nomenclature ($|\mathrm{cb}+\rangle\ldots$) to the eigenstates. Although some authors use a nomenclature that relates these states with the spin-orbit coupled kinetic moment, or with the representations of the symmetry group for the zinc-blende, we have preferred to relate it more straightforwardly with the usual name of the band they belong to. The plus and minus signs refer to the sign of total value of the spin-orbit combined kinetic moment component in an arbitrary axis. They also refer to the energy splitting under a magnetic field along such an axis: those with the + sign move up and those with the − sign move down. Referring to these states we can also call them up and down states respectively.

7.3 The Hamiltonian Matrix Outside the Gamma Point

In this section, we use the k·p Luttinger-Kohn (LK) Hamiltonian [3] with the Pikus-Bir (PB) modifications [4, 5] to include the effect of strained lattices caused by the inclusion of the nanostructures. The explanation presented here follows Ref. [6] for an 8-band (8B) matrix. The matrix is the development of the Hamiltonian in a $|0, v, \pm, \boldsymbol{k}\rangle = |0, v, \pm\rangle e^{i\boldsymbol{k}\cdot\boldsymbol{r}}/\sqrt{\Omega}$ basis where the $|0, v, \pm\rangle$ states are the eigenfunctions (eigenstates) represented in Eq. (7.9). Only matrix elements linking states with the same $\boldsymbol{k}$ are non-zero so that Hamiltonian may be represented as a matrix of dimension eight in which each element is a function of $\boldsymbol{k}$.

The purpose of this chapter is not to discuss the origin of the Hamiltonian to be used, but instead to illustrate how to use it. By ordering the matrix elements as $|cb+\rangle, |hh+\rangle, |lh+\rangle, |so+\rangle, |cb-\rangle, |hh-\rangle, |lh-\rangle, |so-\rangle$, the eight-dimensional matrix may be divided into four blocks.

$$(H) = \begin{pmatrix} (H_{uu}) & (H_{ul}) \\ (H_{lu}) & (H_{ll}) \end{pmatrix} \tag{7.10}$$

The interesting aspect of this block separation is that is $(H_{lu}) = (H_{ul})^{+}$ (hermitical conjugate) and $(H_{ll}) = (H_{uu})^{*}$ (complex conjugate); therefore, only two of the four matrices have to be found. Other advantages will be explained later.

Each one of the block matrices may be considered the sum of a kinetic matrix (the LK part), that applies to non strained materials, and a strained material matrix (the PB part). Auxiliary functions are defined to write the matrix elements. For the different materials they are function of a set of parameters that can be found in the literature [7]. For the InAs and the GaAs they are in Table 7.1 [8].

As discussed in previous chapters, we take the parameters of the QD material, which is InAs in this case. In this chapter some of the preceding parameters are not used directly; the modified Luttinger parameters are

$$\begin{aligned} \gamma_1 &= \gamma_1^L - \frac{E_p}{3E_g + \Delta} \\ \gamma_2 &= \gamma_2^L - \frac{1}{2}\frac{E_p}{3E_g + \Delta} \\ \gamma_3 &= \gamma_3^L - \frac{1}{2}\frac{E_p}{3E_g + \Delta} \end{aligned} \tag{7.11}$$

The parameter B appearing in Fig. 3.1 is, in [8], called the coupling between the conduction and valence bands, and is related to E_p in Table 7.1 by

Table 7.1 Material parameters [8]; γ_{cb} is from [6]

Parameter	InAs	GaAs
γ_{cb}	−1.20	
γ_1^L	19.67	6.85
γ_2^L	8.37	2.1
γ_3^L	9.29	2.9
E_g (eV)	0.418	1.519
Δ (eV)	0.38	0.33
E_p (eV)	22.2	25.7
a_c (eV)	−6.66	−8.6
a_v (eV)	0.66	0.7
b_v (eV)	−1.8	−2.0
d_v (eV)	−3.6	−5.4
l_c (nm)	0.60583	0.56532
B (eV) @ d = 1 nm	0.920	

Table 7.2 Initial inputs calculated with the material parameters in Table 7.1 and values obtained by fitting

Parameter	Initial inputs for eight bands	8B unstrained fitting	8B strained fitting	4B strained fitting
γ_{cb}	−1.202	0.75	−11.2	−3.3
γ_1	6.084	3.5	3.5	3.5
γ_2	1.577	0.53	0.53	0.53
γ_3	2.497	0.53	0.53	0.53
B (eV) @ d = 1 nm	0.920	0.920	1.25	1.35
χ_s		0	0.239	0.239
a_c (eV)	−6.66	–	−6.66	−6.66
a_v (eV)	0.66	–	0.66	0.66
b_v (eV)	−1.8	–	0.0091	0.007
d_v (eV)	−3.6	–	−3.6	−3.6

$$B = \frac{\hbar}{ed}\sqrt{\frac{eE_p}{2m_0}} \tag{7.12}$$

d is an arbitrary normalizing length; usually we multiply the wavevectors in our calculations by it (our calculation variables are $\boldsymbol{K} = d\boldsymbol{k}$).

The numerical values of the preceding expressions appear in Table 7.2 (initial inputs) using the InAs values in Table 7.1. For B we have used Eq. (7.12) with the data in Table 7.1. It differs slightly from the one deduced in Chap. 3 [for d = 1 nm it is 0.751 eV to compare with 0.920 eV obtained with Eq. (7.12)].

The parameter γ_{cb} is used in [6], which inspires this presentation, but not in [8], from where the parameters are extracted. We have set in Table 7.1 the value deduced from [6], although this and other parameters will be fitted later.

The auxiliary functions corresponding to the LK part (also called kinetic) are

$$\begin{aligned}
O_{LK} &= \gamma_c \frac{\hbar^2}{2m_0ed^2}\left(K_x^2 + K_y^2 + K_z^2\right)\\
P_{LK} &= \gamma_1 \frac{\hbar^2}{2m_0ed^2}\left(K_x^2 + K_y^2 + K_z^2\right)\\
Q_{LK} &= \gamma_2 \frac{\hbar^2}{2m_0ed^2}\left(K_x^2 + K_y^2 - 2K_z^2\right)\\
R_{LK} &= \sqrt{3}\frac{\hbar^2}{2m_0ed^2}\left[\gamma_2\left(K_x^2 + K_y^2\right) - 2i\gamma_3 K_x K_y\right]\\
S_{LK} &= \sqrt{6}\frac{\hbar^2}{2m_0ed^2}\gamma_3(K_x - iK_y)K_z\\
T_{LK} &= B(K_x + iK_y)/\sqrt{6}\\
U_{LK} &= BK_z/\sqrt{3}
\end{aligned} \tag{7.13}$$

The LK upper matrix block elements can now be written as

$$(H_{LK,uu}(\boldsymbol{K})) = \begin{pmatrix} E_c + O_{LK} & -\sqrt{3}T_{LK} & \sqrt{2}U_{LK} & -U_{LK} \\ -\sqrt{3}T_{LK}^* & E_v - P_{LK} - Q_{LK} & \sqrt{2}S_{LK} & -S_{LK} \\ \sqrt{2}U_{LK}^* & \sqrt{2}S_{LK}^* & E_v - P_{LK} + Q_{LK} & -\sqrt{2}Q_{LK} \\ -U_{LK}^* & -S_{LK}^* & -\sqrt{2}Q_{LK}^* & E_v - P_{LK} - \Delta \end{pmatrix} \quad (7.14)$$

where the notation makes it explicit that the terms of the matrix are functions of the normalized wavevector $\boldsymbol{K}$. No energy origin is assumed in this matrix. Actually, the conduction band edge will be permitted to vary in the QD and in the barrier material. In certain calculations, it may be simplifying to put the energy origin in the valence band edge, so leading to $E_c = E_g$ and $E_v = 0$.

The non diagonal matrix $(H_{LK,ul})$ is

$$(H_{LK,ul}(\boldsymbol{K})) = \begin{pmatrix} 0 & 0 & -T_{LK}^* & -\sqrt{2}T_{LK} \\ 0 & 0 & R_{LK} & -\sqrt{2}R_{LK} \\ T_{LK}^* & R_{LK} & 0 & \sqrt{3}S_{LK} \\ -U_{LK}^* & -S_{LK}^* & -\sqrt{3}S_{LK} & 0 \end{pmatrix} \quad (7.15)$$

The auxiliary functions for the PB part, corresponding to the strained material, are related to the deformation tensor.

$$\begin{bmatrix} \varepsilon_{xx} & \varepsilon_{xy} & \varepsilon_{xz} \\ \varepsilon_{yx} & \varepsilon_{yy} & \varepsilon_{yz} \\ \varepsilon_{zx} & \varepsilon_{zy} & \varepsilon_{zz} \end{bmatrix} = \begin{bmatrix} \frac{\partial u_x}{\partial x} & \frac{1}{2}\left(\frac{\partial u_x}{\partial y} + \frac{\partial u_y}{\partial x}\right) & \frac{1}{2}\left(\frac{\partial u_x}{\partial z} + \frac{\partial u_z}{\partial x}\right) \\ \frac{1}{2}\left(\frac{\partial u_y}{\partial x} + \frac{\partial u_x}{\partial y}\right) & \frac{\partial u_y}{\partial y} & \frac{1}{2}\left(\frac{\partial u_y}{\partial z} + \frac{\partial u_z}{\partial y}\right) \\ \frac{1}{2}\left(\frac{\partial u_z}{\partial x} + \frac{\partial u_x}{\partial z}\right) & \frac{1}{2}\left(\frac{\partial u_z}{\partial y} + \frac{\partial u_y}{\partial z}\right) & \frac{\partial u_z}{\partial z} \end{bmatrix} \quad (7.16)$$

where u are the displacements of the points (e.g. the atoms' nuclei) with respect to their positions of equilibrium.

The functions are

$$\begin{aligned} O_{PB} &= a_c\left(\varepsilon_{xx} + \varepsilon_{yy} + \varepsilon_{zz}\right) \\ P_{PB} &= -a_v\left(\varepsilon_{xx} + \varepsilon_{yy} + \varepsilon_{zz}\right) \\ Q_{PB} &= -b_v\left(\varepsilon_{xx} + \varepsilon_{yy} + \varepsilon_{zz}\right) \\ R_{PB} &= -\frac{\sqrt{3}}{2}b_v(\varepsilon_{xx} - \varepsilon_{yy}) + id_v\varepsilon_{xy} \\ S_{PB} &= -\frac{d_v}{\sqrt{2}}(\varepsilon_{zx} - i\varepsilon_{yz}) \\ T_{PB} &= \frac{B}{\sqrt{6}}\left[(\varepsilon_{xx} + i\varepsilon_{yx})K_x + (\varepsilon_{xy} + i\varepsilon_{yy})K_y + (\varepsilon_{xz} + i\varepsilon_{yz})K_z\right] \\ U_{PB} &= \frac{B}{\sqrt{3}}K_z/\sqrt{3}(\varepsilon_{zx}K_x + \varepsilon_{zy}K_y + \varepsilon_{zz}K_z) \end{aligned} \quad (7.17)$$

and the upper PB diagonal block is

$$(H_{PB,uu}(\boldsymbol{K})) = \begin{pmatrix} O_{PB} & -\sqrt{3}T_{PB} & \sqrt{2}U_{PB} & -U_{PB} \\ -\sqrt{3}T^*_{PB} & -P_{PB} - Q_{PB} & \sqrt{2}S_{PB} & -S_{PB} \\ \sqrt{2}U^*_{PB} & \sqrt{2}S^*_{PB} & -P_{PB} + Q_{PB} & -\sqrt{2}Q_{PB} \\ -U^*_{PB} & -S^*_{PB} & -\sqrt{2}Q^*_{PB} & -P_{PB} \end{pmatrix} \tag{7.18}$$

The non diagonal bloc is like in Eq. (7.15) but with the subindex *PB*.

As stated before, the Hamiltonian Matrix is the sum of the LK matrix and the PB matrix. For brevity, we call it LKPBH.

7.4 Unstrained Material Parameter Fitting

In this section we develop Ref. [9] further.

7.4.1 Energy Spectra in the Unstrained QD Material

We shall study first the QD material in absence of strain, that is, its situation when it is a bulk material. Only the LK Hamiltonian is to be used in this case. For $\boldsymbol{K} = 0$, using the parameters discussed above, based on Ref. [8], and setting the energy origin at the VB top, the upper diagonal block (in eV) can be written as

$$(H_{k,uu}(0)) = \begin{pmatrix} 0.418 & 0 & 0 & 0 \\ 0 & 0 & 0 & 0 \\ 0 & 0 & 0 & 0 \\ 0 & 0 & 0 & -0.38 \end{pmatrix} \tag{7.19}$$

which reproduces the results of Sect. 7.2 when the values in Table 7.1 are used. The eigenvalues are, 0.418, 0, 0, and −0.38 eV or E_g, E_v, E_v, E_{so}. If the full 8B Hamiltonian is used, the lower diagonal block is the same and all the elements of the non-diagonal blocs are zero; the eigenvalues are the same but each is doubly degenerated. The eigenfunctions are those in Eq. (7.9).

When $\boldsymbol{K} \neq 0$, most of the zero elements disappear. The dispersion curves relating the eigenvalues with $\boldsymbol{K}$ are represented in Fig. 7.2 for the 8B Hamiltonian. The spherical symmetry assumed in Chap. 3 is only approximately reproduced by the Luttinger-Kohn model (as expected from a non-isotropic space, structured as a zincblende) although the direction differences are small and not appreciated by many of the experiments.

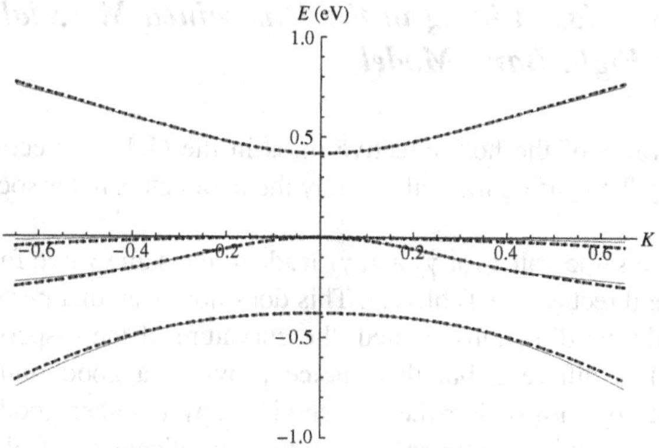

Fig. 7.2 Dispersion curves according to [8] in the (1,1,1) direction (*blue solid line*), in the (1,0,0) or (0,1,0) directions (*red dashed line*) and in the (0,0,1) direction (*black doted line*). Each line is doubly degenerated for the + and − eigenfunctions. The set of curves from up to down correspond to the *cb*, the *hh* the *lh* and the *so* bands

The effective masses implied by these dispersion curves reveal positive effective mass in the conduction band and negative ones in the valence bands. Assuming a parabolic shape, they can be approximately calculated by

$$\frac{m^*}{m_0} = \frac{\hbar^2 K^2}{2d^2 m_0 (E(K) - E(0))} \tag{7.20}$$

being the effective mass independent, in the strictly parabolic case, of the value of K. In our calculations, K is taken as 0.1. Table 7.3 shows the effective masses for the different directions and the experimental values for the unstrained InAs [10]. The calculated values are obtained using Ref. [6] (Tomic et al). We consider that they replicate the experimental data reasonably well (see e.g. [11] for details about the experimental methods), for the conduction and light-hole bands. However it is evident that the calculated heavy-hole effective mass departs from isotropy sensibly and the split-off band is far from the reported measurement.

Table 7.3 Effective masses calculated for unstrained material (data in Table 7.2, "Initial inputs..." column) for the 8B model of the reference, and the experimental values

Eff. mass	Tomic et al. [6]			8B-fitted			Experimental unstrained [10]
Direction	(1,1,1)	(1,0,0)	(0,0,1)	(1,1,1)	(1,0,0)	(0,0,1)	Any
m_{cb}	0.023	0.023	0.023	0.023	0.023	0.023	0.023
m_{hh}	−0.917	−0.341	−0.341	−0.410	−0.410	−0.410	−0.41
m_{lh}	−0.023	−0.024	−0.024	−0.026	−0.026	−0.026	−0.026
m_{so}	−0.064	−0.064	−0.064	−0.077	−0.077	−0.077	−0.16

7.4.2 *Effective Mass Fitting in the Unstrained Material for the Eight Band Model*

The excessive value of the hole effective mass in the (1,1,1) direction suggests a parameter fitting. This fitting must affect only the parameters not associated with the strain.

The use of the same value for γ_2 and γ_3 leads to the same value for the effective masses in all the directions in Table 7.3. This does not mean that perfect isotropy is achieved: outside the directions studied, the curvature of the dispersion functions (effective mass) is different, but this choice provides a good pinning to avoid excessive anisotropy in m_{hh}. For the rest, the isotropy is rather good.

Variations of γ_c and γ_2 produce very minor modifications of all the effective masses in the (1,1,1) direction, although the former can be used for fine trimming of the m_{cb}. This is done as a first step of the fitting process and the result is in Table 7.2, "8B unstrained fitting" column.

Figure 7.3 shows the variation of all the effective masses with $\gamma_2 = \gamma_3$ in the (1,1,1) direction. Since only m_{hh} is strongly variable, it can be adjusted leaving the rest of the effective masses unchanged.

A fine fitting m_{lh} can be achieved by varying γ_1. This will not affect m_{cb} but will modify m_{hh} strongly. Therefore the fitting of m_{lh} and m_{hh} has to be done with γ_1 and $\gamma_2 = \gamma_3$ simultaneously. The resulting parameters are in Table 7.2, "8B unstrained fitting" column and the fitted effective masses appear in Table 7.3 ("8B-fitted" column).

The only effective mass which is not fitted is m_{so}. As the eigenstates of this band are deep into the valence band they do not produce sub-bandgap transitions. Therefore less attention is paid to this band in this work.

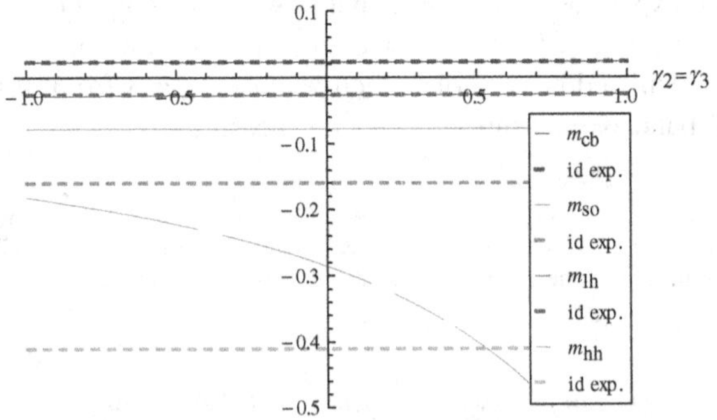

Fig. 7.3 Effective masses in the (1,1,1) direction versus $\gamma_2 = \gamma_3$ when the rest of the parameters are those in Table 7.2 ("8B unstrained fitting" column). *Solid lines* are for the calculated effective mass and *dashed lines* show the corresponding experimental effective mass. The effective mass m_{hh} is fitted for $\gamma_2 = \gamma_3 = 0.53$

7.5 A Simple Strain Assumption for the QD Material

The introduction of a nanostructure in a host of different lattice parameter leads to a deformation of both materials. The displacements of the atoms can be calculated by minimizing the elastic energy. For this, the elastic 4-rank tensor is to be known for the materials involved. Using the displacements in a lattice of calculation points as variables, their positions are calculated to obtain the minimum [12].

The deformations affect significantly the electronic variables. Energy eigenvalues and eigenstates are modified. The electronic energy may be subsequently introduced in the minimization process. In any case, the deformations vary with the position and, consequently, so do the band edges, affecting the offset potentials that in Chap. 3 were taken as square for simplicity. These squared potentials are only possible if very simple deformation tensors are used. Consequently, we assume, as an initial approximation, that the host material is not deformed and that the QD material tends to adopt the lattice parameters of the host material. Under these conditions, taking the origin of coordinates in the center of the QD box used in Chap. 3, the displacements are

$$
\begin{aligned}
u_x(\boldsymbol{r}) &= \chi_s x\left(\frac{l_c(\mathrm{InAs})}{l_c(\mathrm{GaAs})} - 1\right) \\
u_y(\boldsymbol{r}) &= \chi_s y\left(\frac{l_c(\mathrm{InAs})}{l_c(\mathrm{GaAs})} - 1\right) \\
u_z(\boldsymbol{r}) &= \chi_s z\left(\frac{l_c(\mathrm{InAs})}{l_c(\mathrm{GaAs})} - 1\right)
\end{aligned}
\tag{7.21}
$$

where χ_s is a fitting parameter that is one if the QD material takes strictly the lattice constant of the host. The lattice constants in this equation (l_c) may be found in Table 7.1.

By application of Eq. (7.16), the deformation tensor terms are

$$
\begin{aligned}
\varepsilon_{xx} = \varepsilon_{yy} = \varepsilon_{xx} &= \chi_s\left(\frac{l_c(\mathrm{InAs})}{l_c(\mathrm{GaAs})} - 1\right) \\
\varepsilon_{xy} = \varepsilon_{yz} = \varepsilon_{zx} = \varepsilon_{yx} = \varepsilon_{zy} &= \varepsilon_{xz} = 0
\end{aligned}
\tag{7.22}
$$

In reality, the host lattice constant is also modified and strong shear stresses appear in the edges. In all, the potential is not squared; it has peaks in the edges and extends outside the QD. However for the sake of simplicity we have neglected these complexities and adopted square potentials in previous chapters and we shall do the same in this chapter, because its goal is a comparison with the simplified Hamiltonian used so far and approximately verified experimentally.

7.5.1 The Strain-Fitting Coefficient

At the Γ point, the non diagonal bloc matrices are zero so that the 8-band Hamiltonian has the same energy eigenvalues as the upper diagonal matrix. With the strain model described above, the eigenvalue corresponding to the CB bottom is given by

$$E_{CB} = E_g + \left(\frac{l_c(\text{InAs})}{l_c(\text{GaAs})} - 1\right) a_c(\text{InAs})\chi_s \qquad (7.23)$$

whereas E_{VB} is about 10 meV. Thus we can adopt for E_{CB} the experimental bandgap in the strained QD, of 0.734 eV [13]. By taking the parameter values in Table 7.1 (E_g is the unstrained QD material), $\chi_s = 0.238772$.

Notice that the host and QD lattice constants and the unstrained QD bandgap are very well known.

7.5.2 Eight Band Effective Mass Fitting in the Strained Material

The rough approximation adopted for the strain model requires a fitting of the strain-associated parameters. For the strained material, the experimental bandgap is rather different form the bulk InAs, as shown in Table 7.4 ("Chap. 3" column). Also, the effective masses are somewhat different. The following calculations are made using, as a starting point, the fitted parameters for the unstrained material in Table 7.2, "8B unstrained fitting" column.

Among the strain related parameters, the heavy-hole effective mass in the (1,1,1) direction can only be fitted by varying b_v(InAs). As shown in Fig. 7.4, the m_{cb} and m_{so} effective masses do not change and m_{lh} varies slightly. The fitting value is $b_v(\text{InAs}) = -0.0092$ (to be changed in a final fitting).

If the bandgap fitting is to be kept, no fitting of the rest of the effective masses can be found by varying other strain parameters. We are led to revisit the kinetic parameters. γ_{cb}(InAs) affects only the m_{cb} effective mass, but this effective mass is also affected by B(InAs) for which different values have been found depending on the calculation method. The parameter γ_{cb} is considered to be dependent on a parameter to fit m_{cb} and is not primarily associated to the Luttinger parameters: it is not a property of the unstrained material. A simultaneous variation of γ_{cb} and B allows for the simultaneous fitting of m_{cb} and m_{lh}. At the end (when m_{cb} and m_{lh} are fitted) b_v(InAs) has to be fitted again. The fitting values are reported in Table 7.2 ("8B strained fitting" column) and the resulting effective masses are in Table 7.4 ("8B-strained & fitted" column). It can be observed that the isotropy is, in general, very good, especially for the m_{cb}, the m_{so} and the m_{lh}; it is less good for the m_{hh}; the non-fitted m_{so} is not far from the experimental value.

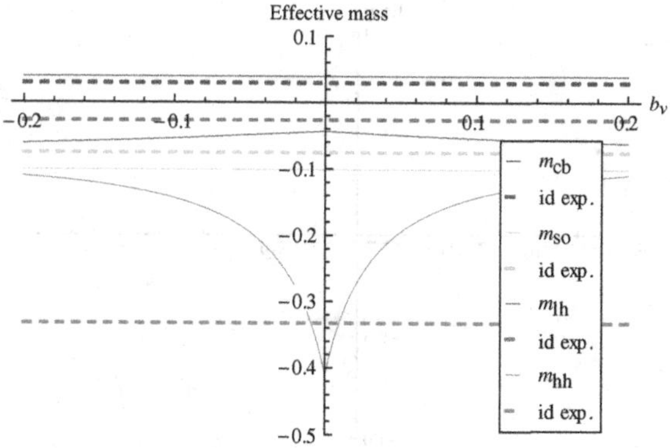

Fig. 7.4 Effective masses versus b_v when the rest of the parameters are those in Table 7.2 (8B unstrained fitting). *Solid lines* are for the calculated effective mass and *dashed lines* show the corresponding experimental effective mass. The effective mass m_{hh} is fitted for $b_v = -0.0092$

7.6 Four-Band Parameter Fitting for the Strained Material

The Empiric k·p Hamiltonian (EKPH) is a four-band k·p model that has been used [13–20] to characterize the absorption behavior in reasonable agreement with measured data in nanostructured solar cells made of zincblende materials. It is presented in Chap. 3 and used in Chaps. 4–6 of this book. We believe it is easy to learn for solid state device physicists and engineers.

In zincblende materials the eight band scheme described in this chapter—the LKPBH—provides a theoretical description of the nanostructure that is widely accepted by theoretical solid state physicists. The main difference between the EKPH and LKPBH is the use of two different sets of eigenvectors as a basis. These basis functions are, in the first case (EKPH), the $|X,\boldsymbol{k}\rangle$, $|Y,\boldsymbol{k}\rangle$, $|Z,\boldsymbol{k}\rangle$, $|S,\boldsymbol{k}\rangle$ functions (without consideration of the spin) and, in the second case (LKPBH), the $|cb\pm,\boldsymbol{k}\rangle$, $|hh\pm,\boldsymbol{k}\rangle$, $|lh\pm,\boldsymbol{k}\rangle$, $|so\pm,\boldsymbol{k}\rangle$ functions (these represent eight functions, depending on the combined angular momentum sign) described earlier in this chapter. However, if we neglect the non-diagonal blocs (that is, we put zero in all their elements) we may restrict our study to the upper diagonal bloc of the Hamiltonian and also restrict the eigenvectors to the four functions $|cb+,\boldsymbol{k}\rangle$, $|hh+,\boldsymbol{k}\rangle$, $|lh+,\boldsymbol{k}\rangle$, $|so+,\boldsymbol{k}\rangle$. Since, in the limit of the validity of the integral factorization rule described in Chap. 2, the absorption properties depend only on the envolvents and not on the GBF, it is irrelevant if these are $|X\rangle$, $|Y\rangle$, $|Z\rangle$, $|S\rangle$ or $|cb+\rangle$, $|hh+\rangle$, $|lh+\rangle$, $|so+\rangle$ (both forming orthonormal sets). In consequence, validating the use of the upper diagonal block only, validates the use of the EKPH model.

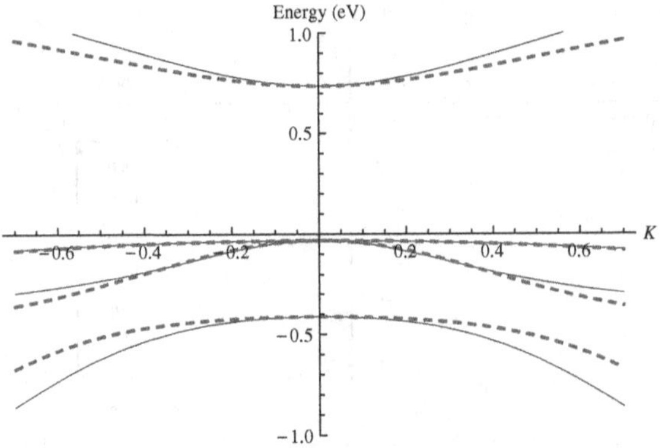

Fig. 7.5 Dispersion curves in the (1,1,1) direction for the 8-band strained InAs (*blue solid line*) and the 4-band upper block (*red dashed line*). Both curves are calculated for the "8B strained fitting" data in column of Table 7.2

We represent in Fig. 7.5 the dispersion curves in the (1,1,1) direction for the 8B and the four-band (4B) models with the parameters of the "8B strained fitting" column of Table 7.2. The role of the non diagonal block is not negligible but is small for small *K*.

Indeed, the agreement will be better if we tune the parameters following the same steps as in Sect. 7.5.2 to fit the effective masses in the (1,1,1) direction. The tuned parameters are reported in Table 7.2 "4B strained fitting" column. The resulting effective masses are in Table 7.4 "4B-strained and fitted" column, together with those for the fitted 8-Band effective masses. It is visible that the 8B model respects better the isotropy observed in the experiments, but not for the m_{hh}, whose isotropy is better for the 4B model.

Table 7.4 Band edges and effective masses with the parameters fitted in Table 7.2 and experimental values used in Chap. 3

Band edges and relative effective mass	8B-strained and fitted			4B-strained and fitted			Chapter 3
E_c (eV)	0.737			0.737			0.737
E_{vhh} (eV)	−0.032			−0.032			0
E_{vlh} (eV)	−0.031			−0.031			0
E_{vso} (eV)	−0.412			−0.412			−0.212
Direction	(1,1,1)	(1,0,0)	(0,0,1)	(1,1,1)	(1,0,0)	(0,0,1)	Any
m_{cb}	0.0294	0.0293	0.0294	0.0294	−0.038	−0.020	0.0294
m_{hh}	−0.333	−0.185	−0.246	−0.333	−0.337	−0.301	−0.333
m_{lh}	−0.027	−0.029	−0.027	−0.027	−0.030	−0.024	−0.027
m_{so}	−0.062	−0.062	−0.062	−0.123	−0.285	−0.058	−0.076

7.7 Eigenvalues and Eigenfunctions in the Nanostructured Material

7.7.1 The Energy Spectrum

When QDs are embedded in a host material, the band offsets confine the electrons around the QDs.

We reproduce here Eq. (3.11) of Chap. 3. This equation is valid for the k·p method in general, without specific reference to the number of bands utilized,

$$\sum_{\nu'} (H_{\boldsymbol{k}})_{\nu,\nu'} \psi_{\nu',\boldsymbol{k}} + \sum_{\boldsymbol{k}'} {}^{\nu}U_{\boldsymbol{k},\boldsymbol{k}'} \psi_{\nu,\boldsymbol{k}'} = E\psi_{\nu,\boldsymbol{k}} \tag{7.24}$$

In this equation, the left superindex ν refers to the band whose eigenfunction is being studied.

We remind the reader that $(H_{\boldsymbol{k}})_{\nu,\nu'}$ are the Hamiltonian elements in the matrix of Eq. (7.10), which depend on $\boldsymbol{k}$, and ${}^{\nu}U_{\boldsymbol{k},\boldsymbol{k}'} = \langle \boldsymbol{k}|{}^{\nu}U(\boldsymbol{r})|\boldsymbol{k}'\rangle$ are the band offset elements corresponding to one of the eight bands ν. Finally $\psi_{\nu,\boldsymbol{k}}$ are the coefficients of the plane wave development of the envelope function corresponding to band ν. The $\psi_{\nu,\boldsymbol{k}}$ are the 8 unknowns of the problem, and they must be solved, for each $\boldsymbol{k}$, with the use of the 8 equations represented in Eq. (7.24). This is a set of homogeneous linear algebraic equations that has the trivial solutions $\psi_{\nu,\boldsymbol{k}} = 0$. Non-trivial solutions appear for certain values of E deduced form the canonic equation, as it is taught in the standard theory of linear equations. These solutions are indeterminate in the sense that one of the values of $\psi_{\nu,\boldsymbol{k}}$ can be arbitrarily fixed, or what is better, the norm of the vector of unknowns (for each $\boldsymbol{k}$) may be made unity. Once the equations are solved for a sufficient number of $\boldsymbol{k}$s we have a plane wave development (equivalent to the Fourier transform) of the eigenfunction (equivalent to the inverse Fourier transform). The use of more or less number of plane waves will give results more or less accurate. Notice that ${}^{\nu}U_{\boldsymbol{k},\boldsymbol{k}'}$ can be calculated with realistic position-variable offsets, as deduced, for instance, from the elastic model.

For details of how to use the Fourier transforms we refer to the reader to Chap. 3. From now on, it will be simplifying to write Eq. (7.24) in more symbolic matrix form,

$$(H_{\boldsymbol{k}})(\psi_{\boldsymbol{k}}) + \sum_{\boldsymbol{k}'} {}^{\nu}U_{\boldsymbol{k},\boldsymbol{k}'}(\psi_{\boldsymbol{k}'}) = E(\psi_{\boldsymbol{k}}) \tag{7.25}$$

Now, we diagonalize the matrix $(H_{\boldsymbol{k}})$ by means of a unitary transformation matrix $(T_{\boldsymbol{k}})$, so that $(H_{d,\boldsymbol{k}}) = (T_{\boldsymbol{k}})(H_{\boldsymbol{k}})(T_{\boldsymbol{k}}^{\dagger})$. The diagonal terms, which are the eigenfunctions of $(H_{\boldsymbol{k}})$, are functions of $\boldsymbol{k}$, approximated by parabolic functions with effective masses, which depend on the orientation as shown in Fig. 7.2. They are presented in Fig. 7.5 (solid line), where the strain has been taken into account, for the direction

(1,1,1). Unlike in Chap. 3—where the transformation matrix ($T_{\boldsymbol{k}}$) is the one formed with the eigenvectors of the ($H_{0,\boldsymbol{k}}$) matrix, which neglects the spin-orbit interaction—now it is the result of diagonalizing the matrix of Eq. (7.10). It is to be observed that $(T_{\boldsymbol{k}})({}^{v}U_{\boldsymbol{k},\boldsymbol{k}'})(T_{\boldsymbol{k}}^{\dagger}) = ({}^{v}U_{\boldsymbol{k},\boldsymbol{k}'})$ because $({}^{v}U_{\boldsymbol{k},\boldsymbol{k}'})$ is a constant with respect to the indices v and v' and the same will happen with its sum with respect to $\boldsymbol{k}'$; also ${}^{v}E$ is not changed by the transformation ($T_{\boldsymbol{k}}$).

With the diagonalized ($H_{d,\boldsymbol{k}}$), the eight equations in Eq. (7.24) are uncoupled and each equation involves one of the variables. These variables are,

$$(\varphi_{\boldsymbol{k}}) = (T_{\boldsymbol{k}})(\psi_{\boldsymbol{k}}) \tag{7.26}$$

We can now adopt the approximation of considering the dispersion function to be isotropic, with the experimental effective mass. As shown in Table 7.4, this is a good approximation with the exception of the *hh* band. We still adopt it. By setting $\boldsymbol{k} \rightarrow -i\nabla$ and using the development of the envelope functions in plane waves (see Chap. 3), that is, making the inverse Fourier transform of the equation $\left(H_{d,\boldsymbol{k}}\right)(\varphi_{\boldsymbol{k}}) = E(\varphi_{\boldsymbol{k}})$ (under the hypothesis of isotropy) we obtain a set of 8 equations like

$$-\frac{\hbar^2}{2m_v^*}\nabla^2\Phi(\boldsymbol{r}) + {}^{v}U(\boldsymbol{r}) = E\Phi(\boldsymbol{r}) \tag{7.27}$$

which is an effective-mass Schrödinger equation. However, the eigenvalue E will correspond to one of the bands, the band v, and only this band will have non-trivial solutions, different from zero, to be obtained with this effective mass equation. Other effective mass equations will be used to find the non-trivial solutions for other bands.

The eigenvalues of this equation form the spectrum of energy for band v. In the case of a box-shaped QD the determination of this spectrum is extremely simple, particularly if we accept the separable approximation already used before in this book. We remind the reader that the eight bands are doubly degenerate and their effective masses are the experimental ones. Then the problem is reduced to the four band problem of the EKPH and the results are exactly the same, doubly degenerate because of the spin. In summary, under the approximations adopted, the EKPH model reproduces exactly the spectrum to be obtained with the more widely accepted 8B LHPBH. In the case of the SOTA cell the energy spectrum is the one in Fig. 3.2.

The same results obtained for the spectrum from the 8B model can be obtained from the 4B model in this chapter, if we disregard the anisotropy and assume isotropic effective masses with the experimental data. Differences will come in the determination of the envelope functions, to be studied next.

7.7.2 The Envelope Functions

The grounds of the k·p method, which are introduced in Chap. 3, are independent of the number of bands under consideration. In this respect the development of a stationary wavefunction in the eight band model is

$$\begin{aligned}\Xi = \Psi_{|cb+\rangle}|cb+\rangle + \Psi_{|hh+\rangle}|hh+\rangle + \Psi_{|lh+\rangle}|lh+\rangle + \Psi_{|so+\rangle}|so+\rangle \\ + \Psi_{|cb-\rangle}|cb-\rangle + \Psi_{|hh-\rangle}|hh-\rangle + \Psi_{|lh-\rangle}|lh-\rangle + \Psi_{|so-\rangle}|so-\rangle\end{aligned} \tag{7.28}$$

The first step for obtaining the envelopes is to develop Φ [of Eq. (7.27)] in plane waves (or, what is the same, find the Fourier transform) so obtaining the eight-components vector $(\varphi_{\boldsymbol{k}})$. Then, in application of Eq. (7.26), and taking into account that the inverse matrix of an unitary matrix (as transformation matrices are) is its hermitic conjugate, we can write

$$(\psi_{\boldsymbol{k}}) = (T_{\boldsymbol{k}}^{+})(\varphi_{\boldsymbol{k}}) \tag{7.29}$$

However, we must remember that, usually, the column vector $(\varphi_{\boldsymbol{k}})$ has a single non-zero element, the one corresponding to the band v to which the eigenvalue E belongs. In consequence,

$$\psi_{v',\boldsymbol{k}} = \sum_{v''}(T_{\boldsymbol{k}}^{+})_{v',v''}\varphi_{v'',\boldsymbol{k}} = (T_{\mathbf{k}}^{+})_{v',v}\varphi_{v,\boldsymbol{k}} = (T_{\mathbf{k}}^{*})_{v,v'}\varphi_{v,\boldsymbol{k}} \tag{7.30}$$

$\varphi_{v,\boldsymbol{k}}$ being the only non-zero element of $(\varphi_{\boldsymbol{k}})$. The last equality comes from expressing that the hermitical conjugate $(T_{\boldsymbol{k}}^{+})$ is the complex conjugate transposed of $(T_{\boldsymbol{k}})$. Thus, the eight elements of the column vector $(\psi_{\boldsymbol{k}})$ are obtained by multiplying the only non-zero element of vector $(\varphi_{\boldsymbol{k}})$ by the complex conjugates of the eight elements of the row corresponding to the row of the non-zero element of $(\psi_{\boldsymbol{k}})$, which in turn, corresponds to a given band (e.g. the *cv+* band or the *hh−* band). Of course this has to be done for each one of the $\boldsymbol{k}$ values (three-component vectors) which are considered necessary for accuracy. We are talking of a band, but remember that in many cases we refer to an energy level, made discrete by the nanostructure, which corresponds to an eigenfunction of a certain band, e.g. of the conduction band.

Once the eight components $\psi_{v',\boldsymbol{k}}$ have been obtained for as many $\boldsymbol{k}$s as deemed necessary, we can calculate the inverse Fourier transform and obtain Ψ_v which are the envolvents to put in Eq. (7.28). Taking this rule into account, the terms of the complex conjugate of matrix $(T_{\boldsymbol{k}})$, can be labeled, for instance, $(T_{\boldsymbol{k}}^{*})_{|hh+\rangle,lh-}$, the first index expressing on which GBF the envelope is projecting the wavefunction and the second index expressing which band it belongs to.

We insist again that for detailed information about how to use the Fourier transforms and the inverse Fourier transforms, the reader is recommended to go to Chap. 3. There, the rules for obtain the envelopes are detailed and they do not differ from those used here, except in the number of bands and in the fact that the transformation matrix may be complex (it is real in Chap. 3 so that the complex conjugation is not explicitly declared).

7.7.2.1 Some Properties of the Diagonalization Matrix

We want to stress some relationships existing among the Fourier transforms of the functions to be used along this chapter. Let us first remember certain theorems of the one dimensional DFT and IDFT. They derive quite straightforwardly from their definitions, which were presented in Chap. 3, Eq. (3.20). We write then again (the DFT in the first line and the IDFT in the second) with the nomenclature somewhat simplified. Remember that only discrete sets of values labeled with n_z and κ_z are taken for z and k in the following equation ($\propto$ = proportional).

$$\begin{aligned}\varphi(k) &\propto \sum_{n_z=-(N-1)/2}^{(N-1)/2} \zeta(z)e^{-ikz} \\ \zeta(z) &\propto \sum_{\kappa_z=-(N-1)/2}^{(N-1)/2} \varphi(k)e^{ikz}\end{aligned} \tag{7.31}$$

(a) The DFT of any real function fulfils $\varphi(k)^* = \varphi(-k)$.
(b) For even $\zeta(z)$, the DFT fulfils $\varphi(k) = \varphi(-k)$; to prove this, it suffices to change the order of the terms. If this relationship is combined with a), it is concluded that $\varphi(k)$ is real.
(c) For odd $\zeta(z)$, the DFT fulfils that $\varphi(k) = -\varphi(-k)$. If this relationship is combined with (a) it is concluded that $\varphi(k)$ is a pure imaginary.

We must now remember that the 3D DFT is obtained by applying the 1D DFT three times. If we start transforming z, then y and finally x, in the first transformation the variables x and y are considered parameters; in the second transformation variables x and k_z are considered parameters; in the third transformation, variables k_y and k_z are considered parameters.

Assuming we adopt the separation of variables described in Chap. 3,

(d) For functions $\Phi(\boldsymbol{r})$ which are even in all the three coordinates, the 3D $\varphi(\boldsymbol{k})$ is real.
(e) For functions $\Phi(\boldsymbol{r})$ which are even in the two coordinates and odd in the other, the 3D $\varphi(\boldsymbol{k})$ is pure imaginary.
(f) For functions $\Phi(\boldsymbol{r})$ which are even in one coordinate and odd in the other two, the 3D $\varphi(\boldsymbol{k})$ is real.
(g) For functions $\Phi(\boldsymbol{r})$ which are even in all the three coordinates, the 3D $\varphi(\boldsymbol{k})$ is pure imaginary.

$$|(T_k)| = \begin{pmatrix} 0.782 & 0.381 & 0.297 & 0.182 & 0 & 0.001 & 0.221 & 0.271 \\ 0.008 & 0.005 & 0.243 & 0.408 & 0.618 & 0.413 & 0.357 & 0.308 \\ 0.066 & 0.534 & 0.326 & 0.293 & 0.034 & 0.120 & 0.095 & 0.703 \\ 0.030 & 0.201 & 0.452 & 0.225 & 0 & 0.587 & 0.597 & 0.038 \\ 0 & 0.001 & 0.221 & 0.271 & 0.782 & 0.381 & 0.297 & 0.182 \\ 0.618 & 0.413 & 0.357 & 0.308 & 0.008 & 0.005 & 0.243 & 0.408 \\ 0.034 & 0.120 & 0.095 & 0.703 & 0.066 & 0.534 & 0.326 & 0.293 \\ 0 & 0.587 & 0.597 & 0.038 & 0.030 & 0.201 & 0.452 & 0.225 \end{pmatrix}$$

Fig. 7.6 Absolute value of the matrix $(T_{(1,1,1)})$ in the point K = (1,1,1). The rows correspond to 1 = cb+, 2 = cb−, 3 = so+, 4 = so−, 5 = lh+, 6 = lh−, 7 = hh+, 8 = hh−. The columns to $1 = |\mathrm{cb}+\rangle, 2 = |\mathrm{hh}+\rangle, 3 = |\mathrm{lh}+\rangle, 4 = |\mathrm{so}+\rangle, 5 = |\mathrm{cb}-\rangle, 6 = |\mathrm{hh}-\rangle, 7 = |\mathrm{lh}+\rangle, 8 = |\mathrm{so}-\rangle$

In summary, $\varphi(\boldsymbol{k})$ is pure imaginary if the number of odd 1D functions is odd. Otherwise it is real.

The matrix (T) has some interesting properties. We present in Fig. 7.6 the absolute value of $(T_{(1,1,1)})$ at the point $\boldsymbol{K} = (1,1,1)$ obtained for $d = 1$ nm. We observe that the two 4D diagonal blocks are equal and the same happens for the two 4D off-diagonal blocs. For example:

$$\begin{aligned} \left|[(T_{\boldsymbol{k}}^*)_{hh+,|lh+\rangle}]\right| = \left|(T_{\boldsymbol{k}}^*)_{2,2}\right| = \left|(T_{\boldsymbol{k}}^*)_{6,6}\right| = \left|[(T_{\boldsymbol{k}}^*)_{hh-,|lh-\rangle}]\right| \\ \left|[(T_{\boldsymbol{k}}^*)_{hh-,|lh+\rangle}]\right| = \left|(T_{\boldsymbol{k}}^*)_{6,2}\right| = \left|(T_{\boldsymbol{k}}^*)7_{2,6}\right| = \left|[(T_{\boldsymbol{k}}^*)_{hh+,|lh-\rangle}]\right| \end{aligned} \tag{7.32}$$

As we have thoroughly verified numerically (no rational proof sought), this also happens for any other value of $\boldsymbol{K}$. Restricting the indices $m \leq 4$ and $n \leq 4$ to the first 4D diagonal block of the matrix this can be stated with generality as

$$\begin{aligned} \left|(T_{\boldsymbol{k}})_{m,n}\right| = \left|(T_{\boldsymbol{k}})_{m+4,n+4}\right| \\ \left|(T_{\boldsymbol{k}})_{m+4,n}\right| = \left|(T_{\boldsymbol{k}})_{m,n+4}\right| \end{aligned} \tag{7.33}$$

7.7.2.2 The Envelope Functions in the Eight Band Model

As we have stated, the eight band dispersion curves are degenerate and the + and −bands correspond to the same dispersion curve. This makes the eigenvalues degenerate forming two dimensional subspaces in the (cb+, cb−), (hh+, hh−). (lh+, lh−) and (so+, so−) couples of bands. Therefore, the mathematical codes to yield the eigenvectors may give any vector of these subspaces, which, indeed, fulfills the eigenvalues equation for a given energy of the subspace. If two eigenvectors are given, they often are not orthogonal and a process of orthogonalization has to be undertaken.[1] Furthermore, in numerical calculations, the eigenvectors are obtained

[1] If $\boldsymbol{a}$ and $\boldsymbol{b}$ are the two non orthogonal eigenvectors forming a vector subspace the vector $\boldsymbol{b}' = \boldsymbol{b} - (\boldsymbol{a} \cdot \boldsymbol{b})\boldsymbol{a}$ belongs to the same subspace and is orthogonal to $\boldsymbol{a}$. It must be subsequently normalized for easy operation.

for any value of $\boldsymbol{k}$ in a way that may appear to be at random by the code user. It may happen that two close $\boldsymbol{k}$ points give eigenvectors which are far away from one another. This mathematical difficulty is related to the fact that it may be physically impossible to determine (in absence of a magnetic field) if a given state belongs to the band + or −, that is, if it is a spin-up or a spin-down state. Nevertheless the eigenvectors obtained are totally valid and can be used to form the diagonalization matrix (T).

To calculate an envelope function we calculate a solution Φ of the TISE Eq. (7.27) and then we obtain the Fourier transform, $\varphi_{\boldsymbol{k}}$. It is the same for the spin-up and spin-down cases. In fact, as said above, it is impossible, in the absence of a magnetic field, to determine if the state is a + or a −band state. As stated in Eq. (7.30), the next step is to multiply this transform by the complex conjugate of the corresponding element of the diagonalization matrix, and finally we must calculate the inverse Fourier transform.

The envelope functions calculated in this way can be denoted, for example, as $\Psi^{(hh+)}_{|lh+\rangle}$; here the $|lh+\rangle$ envelope belongs to an eigenfunction of the band (hh+). The projection of this envelope on its GBF, is

$$\begin{aligned}\left\langle \Psi^{(hh+)}_{|lh+\rangle} \,\middle|\, \Psi^{(hh+)}_{|lh+\rangle} \right\rangle &= \frac{1}{\Omega}\int \left|\Psi^{(hh+)}_{|lh+\rangle}\right|^2 d^3r \propto \int \left|\psi^{(hh+)}_{|lh+\rangle}\right|^2 d^3k \\ &= \int \left|T^*_{hh+,|lh+\rangle}\phi^{(hh+)}\right|^2 d^3k \end{aligned} \tag{7.34}$$

The proportionality sign ($\propto$) is the application of the Parceval's theorem of the Fourier transforms. It is not an equality because ψ are the coefficients of a plane wave development of Ψ, which are proportional but not equal to its Fourier transform. The proportionality factor can be found in Eq. (3.22). Any other envelope function can be written by substituting the subindices as required.

We assume that there are a big number of similar systems (QDs in the nanostructured semiconductor) and we look specifically to one band, e.g. the lh band. Many eigenvectors in this band are spin-up and many others are spin down. Under the optics of the statistical quantum mechanics, in a stationary state they are equally probable [1], with probability 0.5. They constitute a statistical mixture; there is no correlation among them. There is not an eigenstate representing the mixture. It is intuitive to think that the projection on the spin up and spin down modality of each envelope are the same. Let us calculate the projection on GBF $|lh+\rangle$ of an eigenfunction in the band (cb) without specification of the spin. According to Eq. (7.34) it will be

$$\begin{aligned}\left\langle \Psi^{(hh)}_{|lh+\rangle} \,\middle|\, \Psi^{(hh)}_{|lh+\rangle} \right\rangle &\equiv 0.5\left(\left\langle \Psi^{(hh+)}_{|lh+\rangle} \,\middle|\, \Psi^{(hh+)}_{|lh+\rangle} \right\rangle + \left\langle \Psi^{(hh-)}_{|lh+\rangle} \,\middle|\, \Psi^{(hh-)}_{|lh+\rangle} \right\rangle\right) \\ &\propto \int 0.5\left|\phi^{(hh+)}\right|^2\left(\left|T^*_{hh+,|lh+\rangle}\right|^2 + \left|T^*_{hh-,|lh+\rangle}\right|^2\right) d^3k \end{aligned} \tag{7.35}$$

The left hand side of the equalities represents the projection on the GBF, no matter the spin of the state. It can be noticed that $|\phi^{(hh+)}|^2$ has been extracted from the absolute value calculation. This is in application of the theorems (d)–(g) of Sect. 7.7.2 that state that $\phi^{(hh+)}$ is either real or pure imaginary. We can now repeat the procedure for the GBF $|\mathrm{lh}-\rangle$, the result being

$$\begin{aligned}\left\langle \Psi^{(hh)}_{|lh-\rangle} \;\middle|\; \Psi^{(hh)}_{|lh-\rangle}\right\rangle &\equiv 0.5\left(\left\langle \Psi^{(hh+)}_{|lh-\rangle} \;\middle|\; \Psi^{(hh+)}_{|lh-\rangle}\right\rangle + \left\langle \Psi^{(hh-)}_{|lh-\rangle} \;\middle|\; \Psi^{(hh-)}_{|lh-\rangle}\right\rangle\right)\\ &\propto \int 0.5\left|\phi^{(hh-)}\right|^2\left(\left|T^*_{hh+,|lh-\rangle}\right|^2+\left|T^*_{hh-,|lh-\rangle}\right|^2\right)d^3k \qquad (7.36)\end{aligned}$$

The integrand part in parentheses is the same in Eqs. (7.35) and (7.36) as can be verified by summing the equalities in the two lines of Eq. (7.32). Since $\psi^{(hh+)} = \psi^{(hh-)}$, $\left\langle \Psi^{(hh)}_{|lh+\rangle} \;\middle|\; \Psi^{(hh)}_{|lh+\rangle}\right\rangle = \left\langle \Psi^{(hh)}_{|lh-\rangle} \;\middle|\; \Psi^{(hh)}_{|lh-\rangle}\right\rangle$. The average projection of the mixed spin wavefunction on the spin-up and -down GBF is the same, as it was intuitively expected.

7.7.3 The Absorption Coefficient

The absorption coefficient of light by transitions between two states was given in Eq. (2.70) of Chap. 2. This coefficient is proportional to the square of the absolute value of the electric dipole-matrix element $|\langle \Xi|\boldsymbol{r}\cdot\boldsymbol{\varepsilon}|\Xi'\rangle|^2$. By application of the integral factorization rule, Eq. (3.29) of Chap. 3 can be expressed in this chapter as

$$\begin{aligned}\langle \Xi|\boldsymbol{\varepsilon}\cdot\boldsymbol{r}|\Xi'\rangle &\cong \sum_{v,v'}\langle u_{0,v} \mid u_{0,v'}\rangle\langle \Psi_v|\boldsymbol{\varepsilon}\cdot\boldsymbol{r}|\Psi'_{v'}\rangle = \sum_{v}\langle \Psi_v|\boldsymbol{\varepsilon}\cdot\boldsymbol{r}|\Psi'_v\rangle\\ &= \left\langle \Psi_{|cb+\rangle}\middle|\boldsymbol{\varepsilon}\cdot\boldsymbol{r}\middle|\Psi'_{|cb+\rangle}\right\rangle + \left\langle \Psi_{|hh+\rangle}\middle|\boldsymbol{\varepsilon}\cdot\boldsymbol{r}\middle|\Psi'_{|hh+\rangle}\right\rangle\\ &\quad+\left\langle \Psi_{|lk+\rangle}\middle|\boldsymbol{\varepsilon}\cdot\boldsymbol{r}\middle|\Psi'_{|lh+\rangle}\right\rangle + \left\langle \Psi_{|so+\rangle}\middle|\boldsymbol{\varepsilon}\cdot\boldsymbol{r}\middle|\Psi'_{|so+\rangle}\right\rangle\\ &\quad+\left\langle \Psi_{|cb-\rangle}\middle|\boldsymbol{\varepsilon}\cdot\boldsymbol{r}\middle|\Psi'_{|cb-\rangle}\right\rangle + \left\langle \Psi_{|hh-\rangle}\middle|\boldsymbol{\varepsilon}\cdot\boldsymbol{r}\middle|\Psi'_{|hh-\rangle}\right\rangle\\ &\quad+\left\langle \Psi_{|lk-\rangle}\middle|\boldsymbol{\varepsilon}\cdot\boldsymbol{r}\middle|\Psi'_{|lh-\rangle}\right\rangle + \left\langle \Psi_{|so-\rangle}\middle|\boldsymbol{\varepsilon}\cdot\boldsymbol{r}\middle|\Psi'_{|so-\rangle}\right\rangle \qquad (7.37)\end{aligned}$$

We can then follow the rest of Chap. 3, with the exception that in this chapter, there are eight terms, non-spin-degenerated, whereas in Chap. 3 there were four terms, each with a factor 2 due to the spin degeneracy.

Let us first consider the spin degeneracy in more detail. We consider the absorption between a given initial electron state in a lower +band and a upper state in bands + and −. Equation (7.37) can now be applied for the transition to the + and −bands. For example,

$$\begin{aligned}
\left\langle \Xi^{(hh+)} \middle| \boldsymbol{\varepsilon}\cdot\boldsymbol{r} \middle| \Xi^{(cb+)} \right\rangle \cong & \left\langle \Psi^{(hh+)}_{|cb+\rangle} \middle| \boldsymbol{\varepsilon}\cdot\boldsymbol{r} \middle| \Psi^{(cb+)}_{|cb+\rangle} \right\rangle + \left\langle \Psi^{(hh+)}_{|hh+\rangle} \middle| \boldsymbol{\varepsilon}\cdot\boldsymbol{r} \middle| \Psi^{(cb+)}_{|hh+\rangle} \right\rangle \\
& + \left\langle \Psi^{(hh+)}_{|lh+\rangle} \middle| \boldsymbol{\varepsilon}\cdot\boldsymbol{r} \middle| \Psi^{(cb+)}_{|lh+\rangle} \right\rangle + \left\langle \Psi^{(hh+)}_{|so+\rangle} \middle| \boldsymbol{\varepsilon}\cdot\boldsymbol{r} \middle| \Psi^{(cb+)}_{|so+\rangle} \right\rangle \\
& + \left\langle \Psi^{(hh+)}_{|cb-\rangle} \middle| \boldsymbol{\varepsilon}\cdot\boldsymbol{r} \middle| \Psi^{(cb+)}_{|cb-\rangle} \right\rangle + \left\langle \Psi^{(hh+)}_{|hh-\rangle} \middle| \boldsymbol{\varepsilon}\cdot\boldsymbol{r} \middle| \Psi^{(cb+)}_{|hh-\rangle} \right\rangle \\
& + \left\langle \Psi^{(hh+)}_{|lh-\rangle} \middle| \boldsymbol{\varepsilon}\cdot\boldsymbol{r} \middle| \Psi^{(cb+)}_{|lh-\rangle} \right\rangle + \left\langle \Psi^{(hh+)}_{|so-\rangle} \middle| \boldsymbol{\varepsilon}\cdot\boldsymbol{r} \middle| \Psi^{(cb+)}_{|so-\rangle} \right\rangle \\
\left\langle \Xi^{(hh+)} \middle| \boldsymbol{\varepsilon}\cdot\boldsymbol{r} \middle| \Xi^{(cb-)} \right\rangle \cong & \left\langle \Psi^{(hh+)}_{|cb+\rangle} \middle| \boldsymbol{\varepsilon}\cdot\boldsymbol{r} \middle| \Psi^{(cb-)}_{|cb+\rangle} \right\rangle + \left\langle \Psi^{(hh+)}_{|hh+\rangle} \middle| \boldsymbol{\varepsilon}\cdot\boldsymbol{r} \middle| \Psi^{(cb-)}_{|hh+\rangle} \right\rangle \\
& + \left\langle \Psi^{(hh+)}_{|lh+\rangle} \middle| \boldsymbol{\varepsilon}\cdot\boldsymbol{r} \middle| \Psi^{(cb-)}_{|lh+\rangle} \right\rangle + \left\langle \Psi^{(hh+)}_{|so+\rangle} \middle| \boldsymbol{\varepsilon}\cdot\boldsymbol{r} \middle| \Psi^{(cb-)}_{|so+\rangle} \right\rangle \\
& + \left\langle \Psi^{(hh+)}_{|cb-\rangle} \middle| \boldsymbol{\varepsilon}\cdot\boldsymbol{r} \middle| \Psi^{(cb-)}_{|cb-\rangle} \right\rangle + \left\langle \Psi^{(hh+)}_{|hh-\rangle} \middle| \boldsymbol{\varepsilon}\cdot\boldsymbol{r} \middle| \Psi^{(cb-)}_{|hh-\rangle} \right\rangle \\
& + \left\langle \Psi^{(hh+)}_{|lh-\rangle} \middle| \boldsymbol{\varepsilon}\cdot\boldsymbol{r} \middle| \Psi^{(cb-)}_{|lh-\rangle} \right\rangle + \left\langle \Psi^{(hh+)}_{|so-\rangle} \middle| \boldsymbol{\varepsilon}\cdot\boldsymbol{r} \middle| \Psi^{(cb-)}_{|so-\rangle} \right\rangle
\end{aligned} \tag{7.38}$$

The square of the absolute value of these elements of matrix are to be included in the absorption formula, as is done in Chap. 3 (with the necessary consideration concerning polarization).

For the states belonging to the −band, the terms are

$$\begin{aligned}
\left\langle \Xi^{(hh-)} \middle| \boldsymbol{\varepsilon}\cdot\boldsymbol{r} \middle| \Xi^{(cb+)} \right\rangle \cong & \left\langle \Psi^{(hh-)}_{|cb+\rangle} \middle| \boldsymbol{\varepsilon}\cdot\boldsymbol{r} \middle| \Psi^{(cb+)}_{|cb+\rangle} \right\rangle + \left\langle \Psi^{(hh-)}_{|hh+\rangle} \middle| \boldsymbol{\varepsilon}\cdot\boldsymbol{r} \middle| \Psi^{(cb+)}_{|hh+\rangle} \right\rangle \\
& + \left\langle \Psi^{(hh-)}_{|lh+\rangle} \middle| \boldsymbol{\varepsilon}\cdot\boldsymbol{r} \middle| \Psi^{(cb+)}_{|lh+\rangle} \right\rangle + \left\langle \Psi^{(hh-)}_{|so+\rangle} \middle| \boldsymbol{\varepsilon}\cdot\boldsymbol{r} \middle| \Psi^{(cb+)}_{|so+\rangle} \right\rangle \\
& + \left\langle \Psi^{(hh-)}_{|cb-\rangle} \middle| \boldsymbol{\varepsilon}\cdot\boldsymbol{r} \middle| \Psi^{(cb+)}_{|cb-\rangle} \right\rangle + \left\langle \Psi^{(hh-)}_{|hh-\rangle} \middle| \boldsymbol{\varepsilon}\cdot\boldsymbol{r} \middle| \Psi^{(cb+)}_{|hh-\rangle} \right\rangle \\
& + \left\langle \Psi^{(hh-)}_{|lh-\rangle} \middle| \boldsymbol{\varepsilon}\cdot\boldsymbol{r} \middle| \Psi^{(cb+)}_{|lh-\rangle} \right\rangle + \left\langle \Psi^{(hh-)}_{|so-\rangle} \middle| \boldsymbol{\varepsilon}\cdot\boldsymbol{r} \middle| \Psi^{(cb+)}_{|so-\rangle} \right\rangle \\
\left\langle \Xi^{(hh-)} \middle| \boldsymbol{\varepsilon}\cdot\boldsymbol{r} \middle| \Xi^{(cb-)} \right\rangle \cong & \left\langle \Psi^{(hh-)}_{|cb+\rangle} \middle| \boldsymbol{\varepsilon}\cdot\boldsymbol{r} \middle| \Psi^{(cb-)}_{|cb+\rangle} \right\rangle + \left\langle \Psi^{(hh-)}_{|hh+\rangle} \middle| \boldsymbol{\varepsilon}\cdot\boldsymbol{r} \middle| \Psi^{(cb-)}_{|hh+\rangle} \right\rangle \\
& + \left\langle \Psi^{(hh-)}_{|lh+\rangle} \middle| \boldsymbol{\varepsilon}\cdot\boldsymbol{r} \middle| \Psi^{(cb-)}_{|lh+\rangle} \right\rangle + \left\langle \Psi^{(hh-)}_{|so+\rangle} \middle| \boldsymbol{\varepsilon}\cdot\boldsymbol{r} \middle| \Psi^{(cb-)}_{|so+\rangle} \right\rangle \\
& + \left\langle \Psi^{(hh-)}_{|cb-\rangle} \middle| \boldsymbol{\varepsilon}\cdot\boldsymbol{r} \middle| \Psi^{(cb-)}_{|cb-\rangle} \right\rangle + \left\langle \Psi^{(hh-)}_{|hh-\rangle} \middle| \boldsymbol{\varepsilon}\cdot\boldsymbol{r} \middle| \Psi^{(cb-)}_{|hh-\rangle} \right\rangle \\
& + \left\langle \Psi^{(hh-)}_{|lh-\rangle} \middle| \boldsymbol{\varepsilon}\cdot\boldsymbol{r} \middle| \Psi^{(cb-)}_{|lh-\rangle} \right\rangle + \left\langle \Psi^{(hh-)}_{|so-\rangle} \middle| \boldsymbol{\varepsilon}\cdot\boldsymbol{r} \middle| \Psi^{(cb-)}_{|so-\rangle} \right\rangle
\end{aligned} \tag{7.39}$$

All the terms are to be calculated in this case. As the probability of finding the initial states with up and down spins are 0.5, all the absorption terms found in this chapter have to be multiplied by 0.5 and then summed. If, as usual, the number of states counted in the absorption formulae equals the number of QDs, as done in Chap. 3 (each one with degeneracy 2 due to spin) all the four absorption terms are to be considered in this case (without any degeneracy consideration). If the spin 2 degeneracy is still considered, all the terms have to be multiplied by 0.5, as done in the preceding subsection.

7.7.4 Application of the Eight Band Model to Our Exemplary Solar Cell

The application of the 8B LKPBH to QD solar cells is very time consuming, as compared to the EKPH model. We shall explain why.

The calculation of the envelope functions starts with the calculation of eigenfunctions of the TISE equation with the effective mass of the final state, which is *cb* state (it may be an IB or a virtual bound state). Since, for the case of a box shaped QD with squared potential, the solution is analytic and well known, no time is needed for its determination (it is known beforehand). However the treatment used requires the numeric calculation of the eigenfunction in the nodes of the calculation lattice. This lattice, in our calculations, consists of 41 × 41 × 41 cells each of $1.5 \times 1.5 \times 1.5\ \text{nm}^3$. The total number of nodes for calculation is 68921. Nevertheless, the time for this calculation, for example, for a final state, is in the order of one second, using for it Mathematica and a laptop (Core i5). The same can be said for an initial state.

This calculation is followed by a calculation of the Discrete Fourier Transform in 68921 nodes of the final state and its multiplication by the diagonalization matrix element, which should have been calculated previously (as we see later, this time is counted apart). Then, the same is done for all the *hh* and *lh* initial states, of which there are in the range of 200. Finally, in the case of vertical illumination, the elements of matrix are calculated for x and y polarization. In the EKPH, the calculation of the set of matrix elements associated to a single final state takes about 5 min. In the LKPBH, the time for preceding calculations is multiplied by 2 because there are 8 envelopes instead of 4, that is, 10 min.

Finally, the all the absorption coefficients to a given final state are summed up to obtain the absorption coefficient of the transitions form the VB to this final state; this process may last about 3 min for the EKPH. With a typical total of 10 final states with transitions within the bandgap, the duration is about $(5 + 3) \times 10 = 80$ min. For the LKPBH there is a factor of 8, due to the double number envelopes in each transition times the double number of initial states in the VB times the double number of final states in the IB/CB, in both cases due to the spin degeneracy; thus the duration for the LKPB Hamiltonian is 640 min. The calculation of the total absorption once the absorption to each final state has been calculated is negligible.

The calculation of the diagonalization elements of matrix in the 68921 nodes (the time said before to be counted apart) is by far the most time-consuming factor. In the case of the EKPH model, each element of matrix is analytic and the calculation takes less than 1 min. However, in the 8B LKPBH method, the eigenvalues must be calculated numerically for each one of the 68921 nodes. This takes about 200 min per element in the same machine. Assuming that calculations use the *lh* and the *hh* bands as initial states, and the *cb* band as final states, the matrix elements to be calculated are 12 (the 4 elements in the *cb*, the *lh* and the *hh* rows) for the EKPH method and 48 for the 8B LKPBH method (the 8 elements *cb*, the *lh* and the *hh* rows, each affected of plus and minus spins, that is in 6 rows); in total, $200 \times 4 = 800$ times more than in the EKPH.

By summing up all the before-indicated times, the calculation of the subbandgap absorption with the EKPH may last about $12 + 80 = 92$ min and with the LKPBH it may last $12 \times 800 + 640 = 10240$ min, that is, about 1½ h for the EKPH and over 170 h for the LKPB Hamiltonian. We consider that the LKPB Hamiltonian does not fulfill the requirement of being adequate for device development with feedback of results, at least with the calculation methods used.

Figure 7.7 shows the absorption coefficients of the exemplary cell (the SOTA cell) calculated with the 8B LKPB model described above. All the *cb* states have been considered empty except the *cb*(1,1,1) level, which is considered 80% empty due to the partial filling with the CB dopants. That is why the black curve does not coincide with the cyan curve at low energies. The cyan curve represents the absorption by the *cb*(1,1,1) level, assuming it totally empty.

Figure 7.8 shows the IQE calculated under full collection from the preceding absorption coefficients together with the measured one and those calculated with the EKPH model and the four band (H_{uu}) described in Sect. 7.8.

The black dotted curve corresponds to the 8B LKPB calculations; it is below the rest of the calculations including the experimental one. It seems surprising that this calculation, which, a priori, was deemed the most accurate, replicates the experimental data the worst. However, it must not be forgotten the approximations adopted concerning the strain. In addition, the EKPH has a fitting parameter, which is the Givens rotation, so it is not surprising it can lead to a better fitting. In any case, the 8B LKPB model is also within the order of magnitude of the real absorption, stressing the semi quantitative nature of our approximations.

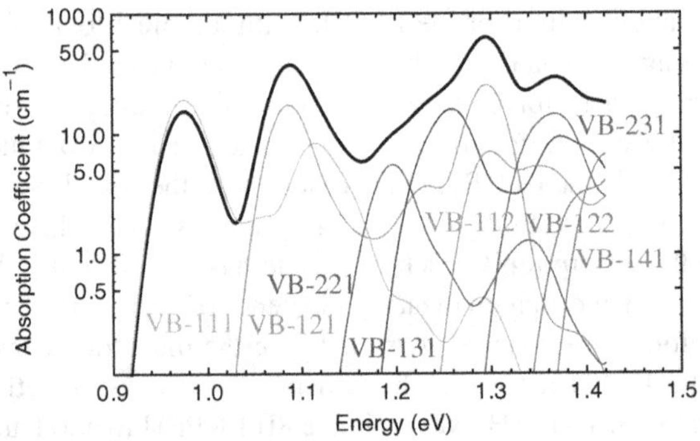

Fig. 7.7 Sub-bandgap absorption calculated with the eight band LKPBH (*thick black line*) and contribution of the transitions from all the bound states in the *hh* and *lh* bands to the *cb* states (including the IB states). In the legends, each *curve* is labeled with the quantum numbers of the *cb* final state. Each absorption curve includes all spin combinations

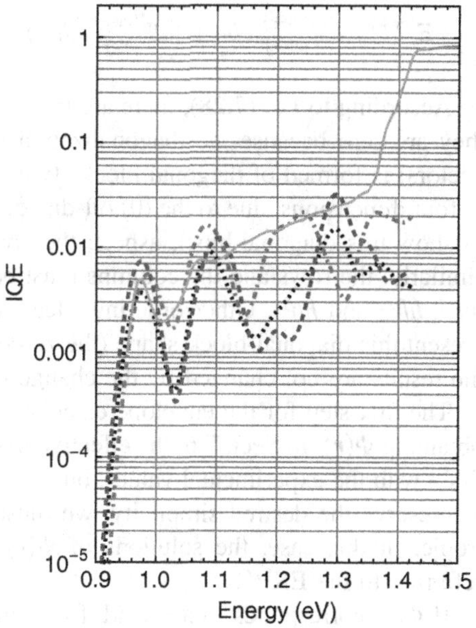

Fig. 7.8 Measured internal quantum efficiency (*IQE*) of the SOTA cells (*green solid line*) as compared with the EKPH calculation for 0.2 Givens rotation (*blue dot-dashed line*), the four band LKPB (H_{uu}) (*red dashed line*) calculation and the eight band LKPB (*black dotted line*) calculation. In all the calculations, the fundamental IB $|cb111\rangle$ state is assumed prefilled to 20 % with doping

7.8 The Four Band Approximation of the Luttinger-Kohn-Pikus-Bir Hamiltonian

The EKPH is a four band Hamiltonian allowing for fast calculations. Along this chapter, and in particular in Sect. 7.6, four band matrices have been examined in the framework of the LHPBH. We are going to examine now what the envelopes and the absorption coefficients for the four bands are like. This four band LKPBH is in fact an 8B Hamiltonian whose off diagonal blocks are set zero, that is,

$$(H) = \begin{pmatrix} (H_{uu}) & (0) \\ (0) & (H^*_{uu}) \end{pmatrix} \tag{7.40}$$

7.8.1 Envelope Functions in the Four Band Approximation

This subsection mainly follows Ref. [21].

Let (ψ_k) be a solution when the Hamiltonian of Eq. (7.40) is put in Eq. (7.25). Assuming that $\Phi(\boldsymbol{r})$ is one of the functions corresponding to a +band, e.g. the cb+, the eigenfunction will be

$$\Xi^{(cv+)} = \Psi^{(cv+)}_{|cb+\rangle}|cb+\rangle + \Psi^{(cv+)}_{|hh+\rangle}|hh+\rangle + \Psi^{(cv+)}_{|lh+\rangle}|lh+\rangle + \Psi^{(cv+)}_{|so+\rangle}|so+\rangle \tag{7.41}$$

According to Eq. (7.28), there are four more envelopes for the labeled GBF, but they are zero because the diagonalization matrix (which is made with the eigenvectors) is formed of diagonal blocks (with zeroes outside these blocks), each one of four dimensions, due to the (0) off-diagonal matrices of (*H*). It is to be noted that, to show this diagonal-block aspect, the arrows and the columns must be ordered similarly: the rows and the columns must be respectively *cb*+, *hh*+, *lh*+, *so*+, *cb*−, *hh*−, *hh*− and *hh*−. Other ordering,[2] leading to matrices called conjugate, do not present this diagonal-block shape (the zeroes are located otherwise) but, of course, the results are not changed by the change of order of the basis vectors.

The first step for the envelope calculations is the solution of the TISE equation obtaining $\Phi(\boldsymbol{r})$. In Sect. 7.6, the effective mass in the (1,1,1) direction has been well fitted with the experimental values, but the isotropy is not very good. Nevertheless, to preserve the desired simplicity, we must assume that the effective mass is isotropic. In this case, the solutions of $\Phi(\boldsymbol{r})$ do not differ from those in Chap. 3, referring to the EKPH.

If $\Phi(\boldsymbol{r})$ corresponds to a −band, for example, to the cb−, the eigenfunction will be

$$\Xi^{(cv-)} = \Psi^{(cv-)}_{|cb-\rangle}|cb-\rangle + \Psi^{(cv-)}_{|hh-\rangle}|hh-\rangle + \Psi^{(cv-)}_{|lh-\rangle}|lh-\rangle + \Psi^{(cv-)}_{|so-\rangle}|so-\rangle \tag{7.42}$$

containing only–labeled GBFs because the envelopes of the positive-labeled GBF are zero for the reasons explained above.

If the envelopes of Eq. (7.41) have been calculated, it is not necessary to calculate those in (7.42). In effect, since the Hamiltonian upper block (H_{uu}) in Eq. (7.40) is hermitical, the lower block is also hermitical. Both have real eigenvalues which are the same. If (ψ_k) is a solution of (H_{uu}) when this Hamiltonian is put in Eq. (7.25), then when the Hamiltonian is the conjugate complex, corresponding to the lower block, the solutions, which are the eigenvectors, are also the conjugate complex, and so is the diagonalization matrix formed by them. This means that the elements $T(\boldsymbol{k})$ for the upper and lower blocks are complex conjugates. Let us call them $T(\boldsymbol{k})$ and $T(\boldsymbol{k})^*$ (the subindices are dropped out) for the upper and lower bocks (or for + and −bands) respectively.

---

[2] The order for the rows we use is that of Mathematica© which is in the order of decreasing absolute value of the eigenvalues to which each eigenvector corresponds: first row, *cb*; second row *so*; third row, *lh*; fourth row, *hh*. The order given by Mathematica© do not keep the aforementioned relation with the bands for higher **k**s, but we have corrected it in our calculations to keep the rows properly associated to the bands. The columns, corresponding to the envelopes, is given in the usual order along this Chapter [that of the +bands in Eq. (7.9)]: first column, |cb⟩; second column, |hh⟩; third column, |lh⟩; fourth column, |so⟩.

A +band envelope $\Psi^{+}(\boldsymbol{r})$ is the IDFT of $T(\boldsymbol{k})\varphi(\boldsymbol{k})$. The corresponding −band envelope $\Psi^{-}(\boldsymbol{r})$ is the IDFT of $T(\boldsymbol{k})^{*}\varphi(\boldsymbol{k})$. In accordance with the statements (a)–(g), of Sect. 7.7.2, we can consider two cases:

(h) When $\varphi(\boldsymbol{k})$ is real, $\Psi^{-}(\boldsymbol{r}) = \Psi^{+}(-\boldsymbol{r})^{*}$
(i) When $\varphi(\boldsymbol{k})$ is pure imaginary, $\Psi^{-}(\boldsymbol{r}) = i\Psi^{+}(-\boldsymbol{r})^{*}$

In summary, the envelopes in Eq. (7.42) are the origin-reflected complex conjugate of those in (7.41) times i if the number of odd 1D functions (in the separation of variables solution of Φ) is odd. In all the cases, $\left|\Psi^{(cv+)}_{|cb+\rangle}(\boldsymbol{r})\right| = \left|\Psi^{(cv-)}_{|cb-\rangle}(-\boldsymbol{r})\right|$, and the same for other envelopes. Furthermore, integral relations of the type $\left\langle \Psi^{(+)}_{|cb+\rangle} \middle| \Psi^{(+)}_{|cb+\rangle} \right\rangle = \left\langle \Psi^{(-)}_{|cb-\rangle} \middle| \Psi^{(-)}_{|cb-\rangle} \right\rangle$ (the integral extended to all the space is invariant with the reflection), representing the projection of the wavefunction in the GBFs $|cb+\rangle$ and $|cb-\rangle$, are also fulfilled.

7.8.2 Absorption Coefficient in the Four Band Approximation

If the H_{uu} four-band approximation is followed, the matrix element in Eq. (7.37) is written differently for transitions between +band and −band solutions. For a $|+\rangle \rightarrow |+\rangle$ transition it is, e.g.

$$\begin{aligned}\left\langle \Xi^{(hh+)} \middle| \boldsymbol{\varepsilon}\cdot\boldsymbol{r} \middle| \Xi^{(cb+)} \right\rangle \cong & \left\langle \Psi^{(hh+)}_{|cb+\rangle} \middle| \boldsymbol{\varepsilon}\cdot\boldsymbol{r} \middle| \Psi^{(cb+)}_{|cb+\rangle} \right\rangle + \left\langle \Psi^{(hh+)}_{|hh+\rangle} \middle| \boldsymbol{\varepsilon}\cdot\boldsymbol{r} \middle| \Psi^{(cb+)}_{|hh+\rangle} \right\rangle \\ & + \left\langle \Psi^{(hh+)}_{|lh+\rangle} \middle| \boldsymbol{\varepsilon}\cdot\boldsymbol{r} \middle| \Psi^{(cb+)}_{|lh+\rangle} \right\rangle + \left\langle \Psi^{(hh+)}_{|so+\rangle} \middle| \boldsymbol{\varepsilon}\cdot\boldsymbol{r} \middle| \Psi^{(cb+)}_{|so+\rangle} \right\rangle\end{aligned} \tag{7.43}$$

and in $(-) \rightarrow (-)$ transition, the expression is similar with the sign–replacing the signs + in the indices. In the case of $(+) \rightarrow (-)$ or $(-) \rightarrow (+)$ transitions, the matrix elements are zero because in the eight terms that might appear, one of the envolvents is zero. Furthermore, for the $(-) \rightarrow (-)$ transitions, all the envelopes are the reflections of the $(+) \rightarrow (+)$ transitions multiplied by i or not according to the summary rule of at the end of Sect. 7.8.1 and by a minus sign resulting from $\boldsymbol{\varepsilon}\cdot\boldsymbol{r}$. However, the term $\left|\left\langle \Xi^{(hh+)} \middle| \boldsymbol{\varepsilon}\cdot\boldsymbol{r} \middle| \Xi^{(cb+)} \right\rangle\right|^{2}$, which is the term that actually appears in the absorption coefficient, taking into account Eq. (7.33), is the same for the (hh+) → (cb+) and the (hh−) → (cb−) transitions. Of course, the same rule holds for other bands.

This leads exactly to the same result as those in Chap. 3 for the case of the EKPH. Also, in this case, it is enough to calculate only the transitions with one of the spin values and multiply the absorption coefficient so obtained by 2 to take into account the reverse spin.

7.8.3 Comparing Absorption Coefficients Calculated Using the Four Band Approximation and the Empiric *k·p* Hamiltonian Approximation

The (H_{uu}) Hamiltonian of the LHPB model is naturally a four band Hamiltonian that may easily be compared to the EKPH. We remind the reader that the eigenvalues of $(H_{uu})_{\boldsymbol{k}}$ form a dispersion function, that can be considered approximately parabolic for any direction of $\boldsymbol{k}$, characterized by an effective mass. However, for different directions, the effective mass may be different. We have fitted the LKPB effective masses for the (1,1,1) direction to be coincident with the experimental values, which are isotropic. Although full isotropy is not achieved for the different directions, we have accepted them as isotropic. For each $\boldsymbol{k}$, the eigenvectors, once normalized, form a unitary matrix that transforms the (H_{uu}), for this $\boldsymbol{k}$, into a diagonal matrix whose elements are the eigenvectors. Repeating the procedure in Chap. 3, we can now solve the four effective mass equations for the *cb*, the *hh*, *lh* and *so* bands and calculate eigenvectors and eigenfunctions, exactly as was done in Chap. 3. Since isotropy has been accepted as fulfilled, the effective masses have been successfully fitted and the experimental ones have been adopted. If a square potential well is assumed, and the offsets are taken to be the experimental ones, then the Hamiltonian in this chapter is not different to that in Chap. 3. In consequence, the energy spectrum is exactly the same as the one discussed in Chap. 3, with the grounds of Chap. 2.

However, there are differences in the envelope functions because the rows of the diagonalization matrix, which is formed with the eigenvectors of (H_{uu}), is different to those used in Chap. 3, based on the eigenvectors of (H_0).

To form the eigenfunction, these envelopes must be multiplied by the GBF, as indicated in Eqs. (7.37) and (7.43).

We remind the reader that the Hamiltonian used in Chap. 3 was somewhat undetermined due to the (H_0) degeneracy. A Givens rotation was used for a precise determination and tests with different rotation angles were used to adjust the result to the experimental values of the internal quantum efficiency. Now the transformation matrix is the one corresponding to the LKPBH and is totally determined.

We want to stress that the complex conjugation expressed in Eq. (7.30) is to be performed to the eigenvectors of the diagonalization matrix. This was also the case in the EKPH of Chap. 3, but there the eigenvectors were real and this complex-conjugation was unnecessary. However, it is also unnecessary in this case. Not performing the complex conjugation operation is equivalent to use the eigenvectors of the lower diagonal block of the Hamiltonian of Eq. (7.40). However, as explained above, the absolute values of the matrix elements appearing in the absorption coefficients are the same than for the upper block Hamiltonian.

Figure 7.8 shows how close the experimental quantum efficiency is to the calculated results obtained from the fitted 4B LKPBH, simplified by the use of the square potential wells. It is reasonably approximate, taking into account we have

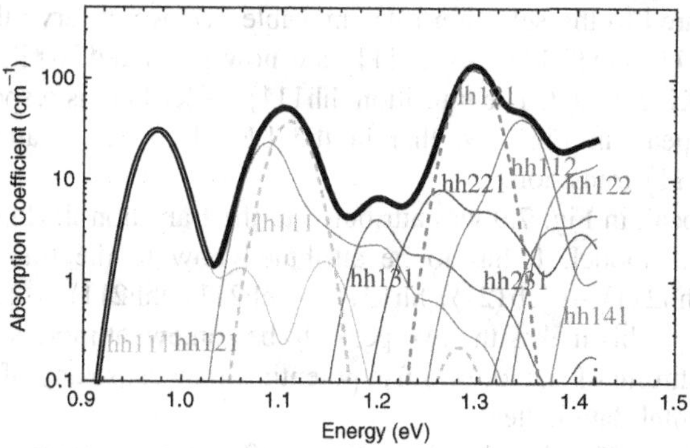

Fig. 7.9 Sub-bandgap absorption calculated with the LKPB (H_{uu}) (*thick black line*) and contribution of the transitions from all the bound states in the *hh* and *lh* bands to the *cb* states (including the IB states). In the legends each curve is labeled with two letters for the initial states and the quantum numbers of the *cb* final state

also neglected the off-diagonal blocks. The approximation with the measured data is not far to the one obtained with the EKPH when the Givens rotation is optimal [14] and both are a little higher than the one obtained with the 8B calculation. All calculations show a more spiky structure. We think this is due to the fact that we have assigned a Gaussian with a constant—energy independent—standard deviation to simulate the Dirac delta peaks smoothed by the different QD sizes. Finally, we have to remember that we do not consider the wetting layer produced in the Stranski-Krastanov growth.

The calculations in the preceding plot are made in the same way as they were made in Chap. 3. The absorption coefficient of all the transitions in the *hh* band *lh* bands to specific levels in the cb are presented in Fig. 7.9.

Indeed, the GBFs are different in this case. They are not |X⟩, |Y⟩, |Z⟩ and |S⟩, but those in Eq. (7.9). However the importance of this is mainly born by the change of diagonalization matrix.

Selection rules have been theoretically obtained for the EKPH model in Chap. 3. This was possible because the diagonalization matrix was analytic and we could determine the table of characters of all the elements. Now this is not the case, but we have been able to extract a rule by looking at all the matrix elements that have been calculated for the drawings in Fig. 7.9. Selection rules are shown in Table 7.5. In contrast no selection rules have been observed in the 8B calculations in Sect. 7.7.4.

Table 7.5 Selection rules for transitions between the *hh* and the *lh* bands and the *cb*. Only non-zero elements are reported

n_x	n_y	n_z	M_x	M_y
0	0	0	Imaginary	Real
1	0	1	Imaginary	Real
1	1	0	Real	Imaginary
0	1	1	Real	Imaginary

0 = equal parity of initial and final states, 1 = different parity

When compared to the selection rules in Table 3.7, we observe that transitions $|\text{hh}111\rangle \rightarrow |\text{cb}111\rangle$ and $|\text{lh}111\rangle \rightarrow |\text{cb}111\rangle$ are now permitted, while they are forbidden in the EKPH model. The transition $|\text{hh}111\rangle \rightarrow |\text{cb}111\rangle$ is responsible for the lowest energy peak in Fig. 7.9, that in the EKPH model is attributed to the $|\text{hh}121\rangle \rightarrow |\text{cb}111\rangle$ transition.

The second peak in Fig. 7.9 was attributed to the transition $|\text{hh}111\rangle \rightarrow |\text{cb}121\rangle$, forbidden in this model. It has to be attributed now to the transition $|\text{hh}121\rangle \rightarrow |\text{cb}121\rangle$(or $|\text{hh}211\rangle \rightarrow |\text{cb}121\rangle, |\text{hh}121\rangle \rightarrow |\text{cb}211\rangle, |\text{hh}211\rangle \rightarrow |\text{cb}211\rangle$, all of the same energy). This makes the two peaks to be somewhat more separated in the (H_{uu}) model. A fitting of size or of offset potentials may cause any of the models to fit the experimental data better.

Looking again at Fig. 7.9, the absorption coefficients represented there that are from transitions from the *hh* and from the *lh* in ending in $|111\rangle$, in $|121\rangle$ (and its degenerate state $|121\rangle$) and in $|221\rangle$ are VB → IB transitions. The rest are sub-bandgap VB → CB transitions caused by the QDs, leading to a shrinkage of the host bandgap. This aspect has already been studied in Chap. 3 and others.

Concerning the time for the calculations, it is the same as that needed for the EKPH model, estimated as 80 min if the time for the calculation of the diagonalization matrix calculation is excluded. This time is in this case 1.75 min per matrix element, to be compared with less than 1 min for the EKPH case, with the same machine and assumptions described in Sect. 7.7.4. In total, the time is $12 \times 1.75 + 80 = 101$ min, or 1 h 41 min. This is slightly more than the 1½ h estimated for the EKPH. This makes the 4B LKPB method equally as acceptable as the EKPH for device calculations and development feedback.

7.9 Concluding Remarks of This Chapter

In the presence of spin, the GBFs of the k·p method are not $|\text{S}\rangle$, $|\text{X}\rangle$, $|\text{Y}\rangle$, $|\text{Z}\rangle$, each one with spin up or down, but different linear combinations of these labeled $|\text{cb}+\rangle$, $|\text{hh}+\rangle$, $|\text{lh}+\rangle$, $|\text{so}+\rangle$, $|\text{cb}-\rangle$, $|\text{hh}-\rangle$, $|\text{lh}-\rangle$, $|\text{so}-\rangle$ in this chapter.

The Luttinger-Kohn Hamiltonian leads to effective masses which are not isotropic, so revealing the GBF structure of the InAs. This anisotropy is not usually found in the experimental data, possibly due to lack of resolution. However, we have found that even without strain, the fitting of the parameters presented in the literature needs to be somewhat modified to fit the calculated values with the experimental values. Due to the simplistic approximation of the strain used in this chapter, modified fitting parameters, are necessary, to obtain the measured values of the effective masses in the presence of strain in InAs/GaAs QD nanostructured material.

This simplistic distribution of the strain, consisting in considering that only the QD is deformed as under hydrostatic pressure, is not realistic, but leads to square potential offsets like those adopted in the EKPH of Chap. 3. The hydrostatic strain is adjusted with a single parameter to the bandgap of the strained QD material.

Realistic strain involves also deformation of the host material and shear stresses in the interfaces. It gives potentials which are not square but that are not very far from square.

The 8B LKPBH leads to four energy dispersion equations. Each is doubly degenerate, but they are not fully isotropic. In the approximation leading to a 4B LKPB Hamiltonian, the dispersion equations are almost the same and also non isotropic. In our calculations, they are assumed isotropic, parabolic and with effective masses equal to the experimental values.

The procedures deduced in Chap. 3 can now be applied to this chapter. In particular, under the parabolic approximation, the energy spectrum for the nano-structured material is exactly the same for the 8B LKPBH, for the 4B LKPBH and for the EKPH of Chap. 3.

The calculation of the envolvents, of the optical matrix elements and of the absorption coefficients is made following the rules of Chap. 3. In the 8B LKPBH, we must consider eight envolvents instead of the four in the EKPH. Also, there are absorption coefficients between up and down states, in contrast with the situation with the EKPH, where only transitions between states of the same spin are permitted. Furthermore, there are no forbidden states caused by the high symmetry of the assumed box-shaped QDs. In spite of all this, the total absorption calculated with the 8B LKPBH is smaller than the experimental one and of the absorption calculated with the EKPH.

In contrast in the 4B LKPBH only transitions between states of the same spin are permitted, like in the EKPH, and rules of selection associated to the QD symmetry are also observed, although different from those in the EKPH.

The most important drawback of the 8B LKPBH is that calculation of the absorption coefficients takes 111 times longer that with the EKPH developed in Chap. 3. The calculation of the full bandgap LKPB Hamiltonian using Mathemetica with a laptop containing a core i5 microprocessor takes over 170 h, which we consider totally inappropriate. With the EKPH, it only takes 1½ h, which we consider reasonable.

However, the 4B LKPBH matrix has a calculation time for the sub-bandgap absorption coefficient that is only 10% more than using the EKPH, and gives quantum efficiency results closer to the measured ones and close to those of the EKPH. We consider this approximation as worth using and appropriate to substitute the EKPH in cases where the uncertainty of the use of the Givens rotation is a nuisance.

References

1. Messiah A (1960) Mécanique Quantique. Dunod, Paris
2. Datta S (1989) Quantum phenomena. Molecular series on solid state devices, vol. 8. Addison Wesley, Reading (Mass)
3. Chuang SL (1995) Physics of optoelectronic devices., Wiley series in pure and applied opticsWiley, New York

4. Pikus GE, Bir GL (1959) Effect of deformation on the energy spectrum and the electrical properties of imperfect germanium and silicon. Sov Phys Solid State 1(1):136–138
5. Pikus GE, Bir GL (1961) Cyclotron and paramagnetic resonance in strained crystals. Phys Rev Lett 6(1):103–105. doi:10.1103/PhysRevLett.6.103
6. Tomic S, Sunderland AG, Bush IJ (2006) Parallel multi-band k center dot p code for electronic structure of zinc blend semiconductor quantum dots. J Mater Chem 16(20):1963–1972
7. Vurgaftman I, Meyer JR, Ram-Mohan LR (2001) Band parameters for III-V compound semiconductors and their alloys. J Appl Phys 89(11):5815–5875
8. Pryor C (1998) Eight-band calculations of strained InAs/GaAs quantum dots compared with one-, four-, and six-band approximations. Phys Rev B 57(12):7190–7195
9. Luque A, Panchak A, Mellor A, Vlasov A, Martí A, Anvreev V (2015) Comparing the Luttinger-Kohn-Pikus-Bir and the empiric k·p Hamiltonians in quantum dot intermediate band solar cells manufactured in zincblende semiconductors. Submitted
10. InAs: Effective masses and density of states. Ioffe Institute Data Basis. http://www.ioffe.ru/SVA/NSM/Semicond/InAs/bandstr.html#Masses
11. Boer KW (1990) Survey of semiconductor physics, vol 1. Van Nostrand Reinhold, New York
12. Pryor C, Pistol ME, Samuelson L (1997) Electronic structure of strained InP/Ga0.51In0.49P quantum dots. Phys Rev B 56(16):10404–10411
13. Luque A, Marti A, Antolín E, Linares PG, Tobías I, Ramiro I, Hernandez E (2011) New Hamiltonian for a better understanding of the quantum dot intermediate band solar cells. Sol Energy Mater Sol Cells 95:2095–2101. doi:10.1016/j.solmat.2011.02.028
14. Luque A, Mellor A, Antolin E, Linares PG, Ramiro I, Tobias I, Marti A (2012) Symmetry considerations in the empirical k·p Hamiltonian for the study of intermediate band solar cells. Sol Energy Mater Sol Cells 103:171–183
15. Luque A, Marti A, Stanley C (2012) Understanding intermediate-band solar cells. Nat Photonics 6(3):146–152. doi:10.1038/nphoton.2012.1
16. Luque A, Mellor A, Ramiro I, Antolín E, Tobías I, Martí A (2013) Interband absorption of photons by extended states in intermediate band solar cells. Sol Energy Mater Sol Cells 115:138–144
17. Luque A, Antolín E, Linares PG, Ramiro I, Mellor A, Tobías I, Martí A (2013) Interband optical absorption in quantum well solar cells. Sol Energy Mater Sol Cells 112:20–26. doi:10.1016/j.solmat.2012.12.045
18. Luque A, Linares PG, Mellor A, Andreev V, Marti A (2013) Some advantages of intermediate band solar cells based on type II quantum dots. Appl Phys Lett 103:123901. doi:10.1063/1.4821580
19. Mellor A, Luque A, Tobias I, Marti A (2013) A numerical study into the influence of quantum dot size on the sub-bandgap interband photocurrent in intermediate band solar cells. AIP Adv 3:022116
20. Mellor A, Luque A, Tobías I, Martí A (2013) Realistic detailed balance study of the quantum efficiency of quantum dot solar cells. Adv Funct Mater. doi:10.1002/adfm.201301513
21. Luque A, Panchak A, Vlasov A, Martí A, Andreev V (2015) Four-band Hamiltonian for fast calculations in intermediate-band solar cells. Submitted